PRAISE FOR

The Physics of Wall Street

"A compelling case for models in economics and an important book for anyone who embraces the scientific method for improving the lot of mankind."

—Michael Brown, former CFO of Microsoft Corporation, past chairman of NASDAQ

"Weatherall's rollicking tale of science and profit has relevance to us all. He goes beyond the 'Frankenstein's monster' cliché to argue that mathematical models are an essential foundation of a saner future."

—William Poundstone, author of *Fortune's Formula*

"[An] enthralling short history of how physics entered Wall Street."

—*Daily Beast*

"This book will lead you to reexamine what you thought you knew about the financial markets, and why it is so important for the economists to actually listen to what the physicists have been trying to tell them."

—Bill Maurer, director of the Institute for Money, Technology, and Financial Inclusion, University of California, Irvine

"Anyone interested in how markets work will appreciate this serious hypothesis."

—*Publishers Weekly*

"Weatherall has a rare talent for making the complex comprehensible, and he puts it to excellent use explaining the role of physics and mathematics in financial markets. This is a book anyone concerned with the unforeseen consequences of financial innovations will want to read."

—Lee Smolin, author of *The Trouble with Physics*

"Beautifully written, with clarity, understanding, and a broad view that is rare in these domains. Even those of us who are unconvinced physics has played an important role in finance will be carried along and learn from this engaging book."

—Stephen M. Stigler, Ernest DeWitt Burton Distinguished Service Professor of Statistics, University of Chicago

"James Weatherall channels the sheer intellectual excitement of unlocking the secrets of nature, whether they relate to fundamental particles or financial markets."

—Hans Halvorson, professor of philosophy, Princeton University

"With *The Physics of Wall Street,* James Weatherall has announced his arrival as one of our leading young science writers. This smart, fast-paced history of ideas—which is packed with vivid portraits of brainiacs famous and obscure and offers a provocative analysis of our current economic woes—should appeal to a broad range of readers, from hard-core science junkies to business folks trying to make sense of modern finance."

—John Horgan, director, Center for Science Writings, Stevens Institute of Technology

"Fascinating history . . . Happily, the author has a gift for making complex concepts clear to lay readers." —*Booklist*

"A lively account of physicists in finance . . . An enjoyable debut appropriate for both specialists and general readers." —*Kirkus Reviews*

The Physics of Wall Street

A BRIEF HISTORY OF PREDICTING THE UNPREDICTABLE

James Owen Weatherall

Mariner Books
Houghton Mifflin Harcourt
BOSTON NEW YORK

First Mariner Books edition 2014

www.hmhco.com

Library of Congress Cataloging-in-Publication Data

Weatherall, James Owen.
The physics of Wall Street : a brief history of predicting the unpredictable / James Owen Weatherall.
p. cm.
Includes bibliographical references and index.
ISBN 978-0-547-31727-4 ISBN 978-0-544-11243-8 (pbk.)
1. Securities—United States. 2. Wall Street (New York, N.Y.) I. Title.
HG4910.W357 2013
332.63′20973—dc23
2012017323

Printed in the United States of America
DOC 10 9 8 7 6 5 4 3 2

To Cailin

Contents

The Physics of Wall Street

Introduction: Of Quants and Other Demons

WARREN BUFFETT ISN'T the best money manager in the world. Neither is George Soros or Bill Gross. The world's best money manager is a man you've probably never heard of—unless you're a physicist, in which case you'd know his name immediately. Jim Simons is co-inventor of a brilliant piece of mathematics called the Chern-Simons 3-form, one of the most important parts of string theory. It's abstract, even abstruse, stuff—some say *too* abstract and speculative—but it has turned Simons into a living legend. He's the kind of scientist whose name is uttered in hushed tones in the physics departments of Harvard and Princeton.

Simons cuts a professorial figure, with thin white hair and a scraggly beard. In his rare public appearances, he usually wears a rumpled shirt and sports jacket—a far cry from the crisp suits and ties worn by most elite traders. He rarely wears socks. His contributions to physics and mathematics are as theoretical as could be, with a focus on classifying the features of complex geometrical shapes. It's hard to even call him a numbers guy—once you reach his level of abstraction, numbers, or anything else that resembles traditional mathematics, are a distant memory. He is not someone you would expect to find wading into the turbulent waters of hedge fund management.

And yet, there he is, the founder of the extraordinarily successful firm Renaissance Technologies. Simons created Renaissance's signature fund in 1988, with another mathematician named James Ax.

They called it Medallion, after the prestigious mathematics prizes that Ax and Simons had won in the sixties and seventies. Over the next decade, the fund earned an unparalleled 2,478.6% return, blowing every other hedge fund in the world out of the water. To give a sense of how extraordinary this is, George Soros's Quantum Fund, the next most successful fund during this time, earned a mere 1,710.1% over the same period. Medallion's success didn't let up in the next decade, either — over the lifetime of the fund, Medallion's returns have averaged almost 40% a year, *after* fees that are twice as high as the industry average. (Compare this to Berkshire Hathaway, which averaged a 20% return from when Buffett turned it into an investment firm in 1967 until 2010.) Today Simons is one of the wealthiest men in the world. According to the 2011 *Forbes* ranking, his net worth is $10.6 billion, a figure that puts Simons's checking account in the same range as that of some high-powered investment firms.

Renaissance employs about two hundred people, mostly at the company's fortresslike headquarters in the Long Island town of East Setauket. A third of them have PhDs — not in finance, but rather, like Simons, in fields like physics, mathematics, and statistics. According to MIT mathematician Isadore Singer, Renaissance is the best physics and mathematics department in the world — which, say Simons and others, is why the firm has excelled. Indeed, Renaissance avoids hiring anyone with even the slightest whiff of Wall Street bona fides. PhDs in finance need not apply; nor should traders who got their start at traditional investment banks or even other hedge funds. The secret to Simons's success has been steering clear of the financial experts. And rightly so. According to the financial experts, people like Simons shouldn't exist. Theoretically speaking, he's done the impossible. He's predicted the unpredictable, and made a fortune doing it.

Hedge funds are supposed to work by creating counterbalanced portfolios. The simplest version of the idea is to buy one asset while simultaneously selling another asset as a kind of insurance policy. Often, one of these assets is what is known as a derivative. Derivatives are contracts based on some other kind of security, such as stocks, bonds, or commodities. For instance, one kind of derivative is called a futures contract. If you buy a futures contract on, say, grain, you are agreeing to buy the grain at some fixed future time, for a price that you settle on now. The value of a grain future depends on the value of grain — if the price of grain goes up, then the value of your grain futures should go up too, since the price of buying grain and holding it for a while should also go up. If grain prices drop, however, you may be stuck with a contract that commits you to paying more than the market price of grain when the futures contract expires. In many cases (though not all), there is no actual grain exchanged when the contract expires; instead, you simply exchange cash corresponding to the discrepancy between the price you agreed to pay and the current market price.

Derivatives have gotten a lot of attention recently, most of it negative. But they aren't new. They have been around for at least four thousand years, as testified by clay tablets found in ancient Mesopotamia (modern-day Iraq) that recorded early futures contracts. The purpose of such contracts is simple: they reduce uncertainty. Suppose that Anum-pisha and Namran-sharur, two sons of Siniddianam, are Sumerian grain farmers. They are trying to decide whether they should plant their fields with barley, or perhaps grow wheat instead. Meanwhile, the priestess Iltani knows that she will require barley next autumn, but she also knows that barley prices can fluctuate unpredictably. On a hot tip from a local merchant, Anum-pisha and Namran-sharur approach Iltani and suggest that she buy a futures contract on their barley; they agree to sell Iltani a fixed amount of barley for a prenegotiated price, after the harvest. That way, Anum-pisha and Namran-sharur can confidently plant barley, since they have already found a buyer. Iltani, meanwhile, knows that she will be able to acquire sufficient amounts of barley at a fixed price. In this case, the derivative reduces the seller's risk of producing the goods in the first place, and at the same time, it shields the purchaser from unexpected variations in price. Of course, there's always a risk that the sons of Siniddianam won't be able to deliver — what if there is a drought or a blight? — in which case they would likely have to buy the grain from someone else and sell it to Iltani at the predetermined rate.

Hedge funds use derivatives in much the same way as ancient Mesopotamians. Buying stock and selling stock market futures is like planting barley and selling barley futures. The futures provide a kind of insurance against the stock losing value.

The hedge funds that came of age in the 2000s, however, did the sons of Siniddianam one better. These funds were run by traders, called quants, who represented a new kind of Wall Street elite. Many had PhDs in finance, with graduate training in state-of-the-art academic theories — never before a prerequisite for work on the Street. Others were outsiders, with backgrounds in fields like mathematics or physics. They came armed with formulas designed to tell them exactly how derivatives prices should be related to the securities on which the derivatives were based. They had some of the fastest, most sophisticated computer systems in the world programmed to solve these equations and to calculate how much risk the funds faced, so that they could keep their portfolios in perfect balance. The funds' strategies were calibrated so that no matter what happened, they would eke out a small profit — with virtually no chance of significant loss. Or at least, that was how they were supposed to work.

But when markets opened on Monday, August 6, 2007, all hell broke loose. The hedge fund portfolios that were designed to make money, no matter what, tanked. The positions that were supposed to go up all went down. Bizarrely, the positions that were supposed to go up if everything else went down *also* went down. Essentially all of the major quant funds were hit, hard. Every strategy they used was suddenly vulnerable, whether in stocks, bonds, currency, or commodities. Millions of dollars started flying out the door.

As the week progressed, the strange crisis worsened. Despite their training and expertise, none of the traders at the quant funds had any idea what was going on. By Wednesday matters were desperate. One large fund at Morgan Stanley, called Process Driven Trading, lost $300 million that day alone. Another fund, Applied Quantitative Research Capital Management, lost $500 million. An enormous, highly secretive Goldman Sachs fund called Global Alpha was down $1.5 *billion* on the month so far. The Dow Jones, meanwhile, went up 150 points, since the stocks that the quant funds had bet *against* all rallied. Something had gone terribly, terribly wrong.

The market shakeup continued through the end of the week. It finally ended over the weekend, when Goldman Sachs stepped in with $3 billion in new capital to stabilize its funds. This helped stop the bleeding long enough for the immediate panic to subside, at least for the rest of August. Soon, though, word of the losses spread to business journalists. A few wrote articles speculating about the cause of what came to be called the quant crisis. Even as Goldman's triage saved the day, however, explanations were difficult to come by. The fund managers went about their business, nervously hoping that the week from hell had been some strange fluke, a squall that had passed. Many recalled a quote from a much earlier physicist. After losing his hat in a market collapse in seventeenth-century England, Isaac Newton despaired: "I can calculate the movements of stars, but not the madness of men."

The quant funds limped their way to the end of the year, hit again in November and December by ghosts of the August disaster. Some, but not all, managed to recover their losses by the end of the year. On average, hedge funds returned about 10% in 2007—less than many other, apparently less sophisticated investments. Jim Simons's Medallion Fund, on the other hand, returned 73.7%. Still, even Medallion had felt the August heat. As 2008 dawned, the quants hoped the worst was behind them. It wasn't.

I began thinking about this book during the fall of 2008. In the year since the quant crisis, the U.S. economy had entered a death spiral, with century-old investment banks like Bear Stearns and Lehman Brothers imploding as markets collapsed. Like many other people, I was captivated by the news of the meltdown. I read about it obsessively. One thing in particular about the coverage jumped out at me. In article after article, I came across the legions of quants: physicists and mathematicians who had come to Wall Street and changed it forever. The implication was clear: physicists on Wall Street were responsible for the collapse. Like Icarus, they had flown too high and fallen. Their waxen wings were "complex mathematical models" imported from physics—tools that promised unlimited wealth in the halls of academia, but that melted when faced with the real-life vicissitudes of Wall Street. Now we were all paying the price.

I was just finishing a PhD in physics and mathematics at the time, and so the idea that physicists were behind the meltdown was especially shocking to me. Sure, I knew people from high school and college who had majored in physics or math and had then gone on to become investment bankers. I had even heard stories of graduate students who had been lured away from academia by the promise of untold riches on Wall Street. But I also knew bankers who had majored in philosophy and English. I suppose I assumed that physics and math majors were appealing to investment banks because they were good with logic and numbers. I never dreamed that physicists were of particular interest because they knew some *physics*.

It felt like a mystery. What could physics have to do with finance? None of the popular accounts of the meltdown had much to say about why physics and physicists had become so important to the world economy, or why anyone would have thought that ideas from physics would have any bearing on markets at all. If anything, the current wisdom — promoted by Nassim Taleb, author of the best-selling book *The Black Swan,* as well as some proponents of behavioral economics — was that using sophisticated models to predict the market was foolish. After all, people were not quarks. But this just left me more confused. Had Wall Street banks like Morgan Stanley and Goldman Sachs been bamboozled by a thousand calculator-wielding con men? The trouble was supposed to be that physicists and other quants were running failing funds worth billions of dollars. But if the whole endeavor was so obviously stupid, why had they been trusted with the money in the first place? Surely someone with some business sense had been convinced that these quants were on to something — and it was *this* part of the story that was getting lost in the press. I wanted to get to the bottom of it.

So I started digging. As a physicist, I figured I would start by tracking down the people who first came up with the idea that physics could be used to understand markets. I wanted to know what the connections between physics and finance were supposed to be, but I also wanted to know how the ideas had taken hold, how physicists had come to be a force on the Street. The story I uncovered took me from turn-of-the-century Paris to government labs during World War II, from blackjack tables in Las Vegas to Yippie communes on the Pacific coast. The connections between physics and modern financial theory — and economics more broadly — run surprisingly deep.

This book tells the story of physicists in finance. The recent crisis is part of the story, but in many ways it's a minor part. This is not a book about the meltdown. There have been many of those, some even focusing on the role that quants played and how the crisis affected them. This book is about something bigger. It is about how the quants came to be, and about how to understand the "complex mathematical models" that have become central to modern finance. Even more importantly, it is a book about the future of finance. It's about why we should look to new ideas from physics and related fields to solve the ongoing economic problems faced by countries around the world. It's a story that should change how we think about economic policy forever.

The history I reveal in this book convinced me — and I hope it will convince you — that physicists and their models are not to blame for our current economic ills. But that doesn't mean we should be complacent about the role of mathematical modeling in finance. Ideas that could have helped avert the recent financial meltdown were developed years before the crisis occurred. (I describe a couple of them in the book.) Yet few banks, hedge funds, or government regulators showed any signs of listening to the physicists whose advances might have made a difference. Even the most sophisticated quant funds were relying on first- or second-generation technology when third- and fourth-generation tools were already available. If we are going to use physics on Wall Street, as we have for thirty years, we need to be deeply sensitive to where our current tools will fail us, and to new tools that can help us improve on what we're doing now. If you think about financial models as the physicists who introduced them thought about them, this would be obvious. After all, there's nothing special about finance — the same kind of careful attention to where current models fail is crucial to all engineering sciences. The danger comes when we use ideas from physics, but we stop thinking like physicists.

There's one shop in New York that remembers its roots. It's Renaissance, the financial management firm that doesn't hire finance experts. The year 2008 hammered a lot of banks and funds. In addition to Bear Stearns and Lehman Brothers, the insurance giant AIG as well as dozens of hedge funds and hundreds of banks either shut down or teetered at the precipice, including quant fund behemoths worth tens of billions of dollars like Citadel Investment Group. Even the traditionalists suffered: Berkshire Hathaway faced its largest loss ever, of about 10% book value per share — while the shares themselves halved in value. But not everyone was a loser for the year. Meanwhile, Jim Simons's Medallion Fund earned 80%, even as the financial industry collapsed around him. The physicists must be doing something right.

CHAPTER 1

Primordial Seeds

LA FIN DE SIÈCLE, LA BELLE EPOQUE. Paris was abuzz with progress. In the west, Gustave Eiffel's new tower — still considered a controversial eyesore by Parisians living in its shadow — shot up over the site of the 1889 World's Fair. In the north, at the foot of Montmartre, a new cabaret called the Moulin Rouge had just opened to such fanfare that the Prince of Wales came over from Britain to see the show. Closer to the center of town, word had begun to spread of certain unexplained accidents at the magnificent and still-new home of the city's opera, the Palais Garnier — accidents that would lead to at least one death when part of a chandelier fell. Rumor had it that a phantom haunted the building.

Just a few blocks east from the Palais Garnier lay the beating heart of the French empire: the Paris Bourse, the capital's principal financial exchange. It was housed in a palace built by Napoleon as a temple to money, the Palais Brongniart. Its outside steps were flanked by statues of its idols: Justice, Commerce, Agriculture, Industry. Majestic neoclassical columns guarded its doors. Inside, its cavernous main hall was large enough to fit hundreds of brokers and staff members. For an hour each day they met beneath ornately carved reliefs and a massive skylight to trade the permanent government bonds, called *rentes,* that had funded France's global ambitions for a century. Imperial and imposing, it was the center of the city at the center of the world.

Or so it would have seemed to Louis Bachelier as he approached it for the first time, in 1892. He was in his early twenties, an orphan from the provinces. He had just arrived in Paris, fresh from his mandatory military service, to resume his education at the University of Paris. He was determined to be a mathematician or a physicist, whatever the odds —and yet, he had a sister and a baby brother to support back home. He had recently sold the family business, which had provided sufficient money for the moment, but it wouldn't last forever. And so, while his classmates threw themselves into their studies, Bachelier would have to work. Fortunately, with a head for numbers and some hard-won business experience, he had been able to secure a position at the Bourse. He assured himself it was only temporary. Finance would have his days, but his nights were saved for physics. Nervously, Bachelier forced himself to walk up the stairs toward the columns of the Bourse.

Inside, it was total bedlam. The Bourse was based on an open outcry system for executing trades: traders and brokers would meet in the main hall of the Palais Brongniart and communicate information about orders to buy or sell by yelling or, when that failed, by using hand signals. The halls were filled with men running back and forth executing trades, transferring contracts and bills, bidding on and selling stocks and *rentes.* Bachelier knew the rudiments of the French financial system, but little more. The Bourse did not seem like the right place for a quiet boy, a mathematician with a scholar's temperament. But there was no turning back. It's just a game, he told himself. Bachelier had always been fascinated by probability theory, the mathematics of chance (and, by extension, gambling). If he could just imagine the French financial markets as a glorified casino, a game whose rules he was about to learn, it might not seem so scary.

He repeated the mantra—*just an elaborate game of chance*—as he pushed forward into the throng.

"Who is this guy?" Paul Samuelson asked himself, for the second time in as many minutes. He was sitting in his office, in the economics department at MIT. The year was 1955, or thereabouts. Laid out in front of him was a half-century-old PhD dissertation, written by a Frenchman whom Samuelson was quite sure he had never heard of. Bachelor, Bacheler. Something like that. He looked at the front of the document again. Louis Bachelier. It didn't ring any bells.

Its author's anonymity notwithstanding, the document open on Samuelson's desk was astounding. Here, fifty-five years previously, Bachelier had laid out the mathematics of financial markets. Samuelson's first thought was that his own work on the subject over the past several years — the work that was supposed to form one of his students' dissertation — had lost its claim to originality. But it was more striking even than that. By 1900, this Bachelier character had apparently worked out much of the mathematics that Samuelson and his students were only now adapting for use in economics — mathematics that Samuelson thought had been developed far more recently, by mathematicians whose names Samuelson knew by heart because they were tied to the concepts they had supposedly invented. Weiner processes. Kolmogorov's equations. Doob's martingales. Samuelson thought this was cutting-edge stuff, twenty years old at the most. But there it all was, in Bachelier's thesis. How come Samuelson had never heard of him?

Samuelson's interest in Bachelier had begun a few days before, when he received a postcard from his friend Leonard "Jimmie" Savage, then a professor of statistics at the University of Chicago. Savage had just finished writing a textbook on probability and statistics and had developed an interest in the history of probability theory along the way. He had been poking around the university library for early-twentieth-century work on probability when he came across a textbook from 1914 that he had never seen before. When he flipped through it, Savage realized that, in addition to some pioneering work on probability, the book had a few chapters dedicated to what the author called "speculation" — literally, probability theory as applied to market speculation. Savage guessed (correctly) that if he had never come across this work before, his friends in economics departments likely hadn't either, and so he sent out a series of postcards asking if anyone knew of Bachelier.

Samuelson had never heard the name. But he was interested in mathematical finance—a field he believed he was in the process of inventing—and so he was curious to see what this Frenchman had done. MIT's mathematics library, despite its enormous holdings, did not have a copy of the obscure 1914 textbook. But Samuelson did find something else by Bachelier that piqued his interest: Bachelier's dissertation, published under the title *A Theory of Speculation.* He checked it out of the library and brought it back to his office.

Bachelier was not, of course, the first person to take a mathematical interest in games of chance. That distinction goes to the Italian Renaissance man Gerolamo Cardano. Born in Milan around the turn of the sixteenth century, Cardano was the most accomplished physician of his day, with popes and kings clamoring for his medical advice. He authored hundreds of essays on topics ranging from medicine to mathematics to mysticism. But his real passion was gambling. He gambled constantly, on dice, cards, and chess—indeed, in his autobiography he admitted to passing years in which he gambled every day. Gambling during the Middle Ages and the Renaissance was built around a rough notion of odds and payoffs, similar to how modern horseraces are constructed. If you were a bookie offering someone a bet, you might advertise odds in the form of a pair of numbers, such as "10 to 1" or "3 to 2," which would reflect how unlikely the thing you were betting on was. (Odds of 10 to 1 would mean that if you bet 1 dollar, or pound, or guilder, and you won, you would receive 10 dollars, pounds, or guilders in winnings, plus your original bet; if you lost, you would lose the dollar, etc.) But these numbers were based largely on a bookie's gut feeling about how the bet would turn out. Cardano believed there was a more rigorous way to understand betting, at least for some simple games. In the spirit of his times, he wanted to bring modern mathematics to bear on his favorite subject.

In 1526, while still in his twenties, Cardano wrote a book that outlined the first attempts at a systematic theory of probability. He focused on games involving dice. His basic insight was that, if one assumed a die was just as likely to land with one face showing as another, one could work out the precise likelihoods of all sorts of combinations occurring, essentially by counting. So, for instance, there are six possible outcomes of rolling a standard die; there is precisely one way in which to yield the number 5. So the mathematical odds of yielding a 5 are 1 in 6 (corresponding to betting odds of 5 to 1). But what about yielding a sum of 10 if you roll two dice? There are $6 \times 6 = 36$ possible outcomes, of which 3 correspond to a sum of 10. So the odds of yielding a sum of 10 are 3 in 36 (corresponding to betting odds of 33 to 3). The calculations seem elementary now, and even in the sixteenth century the results would have been unsurprising — anyone who spent enough time gambling developed an intuitive sense for the odds in dice games — but Cardano was the first person to give a mathematical account of why the odds were what everyone already knew them to be.

Cardano never published his book — after all, why give your best gambling tips away? — but the manuscript was found among his papers when he died and ultimately was published over a century after it was written, in 1663. By that time, others had made independent advances toward a full-fledged theory of probability. The most notable of these came at the behest of another gambler, a French writer who went by the name of Chevalier de Méré (an affectation, as he was not a nobleman). De Méré was interested in a number of questions, the most pressing of which concerned his strategy in a dice game he liked to play. The game involved throwing dice several times in a row. The player would bet on how the rolls would come out. For instance, you might bet that if you rolled a single die four times, you would get a 6 at least one of those times. The received wisdom had it that this was an even bet, that the game came down to pure luck. But de Méré had an instinct that if you bet that a 6 *would* get rolled, and you made this bet every time you played the game, over time you would tend to win slightly more often than you lost. This was the basis for de Méré's gambling strategy, and it had made him a considerable amount of money. However, de Méré also had a second strategy that he thought should be just as good, but for some reason had only given him grief. This second strategy was to always bet that a *double 6* would get rolled at least once, if you rolled two dice twenty-four times. But this strategy didn't seem to work, and de Méré wanted to know why.

As a writer, de Méré was a regular at the Paris salons, fashionable meetings of the French intelligentsia that fell somewhere between cocktail parties and academic conferences. The salons drew educated Parisians of all stripes, including mathematicians. And so, de Méré began to ask the mathematicians he met socially about his problem. No one had an answer, or much interest in looking for one, until de Méré tried his problem out on Blaise Pascal. Pascal had been a child prodigy, working out most of classical geometry on his own by drawing pictures as a child. By his late teens he was a regular at the most important salon, run by a Jesuit priest named Marin Mersenne, and it was here that de Méré and Pascal met. Pascal didn't know the answer, but he was intrigued. In particular, he agreed with de Méré's appraisal that the problem should have a mathematical solution.

Pascal began to work on de Méré's problem. He enlisted the help of another mathematician, Pierre de Fermat. Fermat was a lawyer and polymath, fluent in a half-dozen languages and one of the most capable mathematicians of his day. Fermat lived about four hundred miles south of Paris, in Toulouse, and so Pascal didn't know him directly, but he had heard of him through his connections at Mersenne's salon. Over the course of the year 1654, in a long series of letters, Pascal and Fermat worked out a solution to de Méré's problem. Along the way, they established the foundations of the modern theory of probability.

One of the things that Pascal and Fermat's correspondence produced was a way of precisely calculating the odds of winning dice bets of the sort that gave de Méré trouble. (Cardano's system also accounted for this kind of dice game, but no one knew about it when de Méré became interested in these questions.) They were able to show that de Méré's first strategy was good because the chance that you would roll a 6 if you rolled a die four times was slightly better than 50% — more like 51.7747%. De Méré's second strategy, though, wasn't so great because the chance that you would roll a pair of 6s if you rolled two dice twenty-four times was only about 49.14%, less than 50%. This meant that the second strategy was slightly less likely to win than to lose, whereas de Méré's first strategy was slightly more likely to win. De Méré was thrilled to incorporate the insights of the two great mathematicians, and from then on he stuck with his first strategy.

The interpretation of Pascal and Fermat's argument was obvious, at least from de Méré's perspective. But what do these numbers really mean? Most people have a good intuitive idea of what it means for an event to have a given probability, but there's actually a deep philosophical question at stake. Suppose I say that the odds of getting heads when I flip a coin are 50%. Roughly, this means that if I flip a coin over and over again, I will get heads about half the time. But it doesn't mean I am guaranteed to get heads exactly half the time. If I flip a coin 100 times, I might get heads 51 times, or 75 times, or all 100 times. Any number of heads is possible. So why should de Méré have paid any attention to Pascal and Fermat's calculations? They didn't guarantee that even his first strategy would be successful; de Méré could go the rest of his life betting that a 6 would show up every time someone rolled a die four times in a row and never win again, despite the probability calculation. This might sound outlandish, but nothing in the theory of probability (or physics) rules it out.

So what do probabilities tell us, if they don't guarantee anything about how often something is going to happen? If de Méré had thought to ask this question, he would have had to wait a long time for an answer. Half a century, in fact. The first person who figured out how to think about the relationship between probabilities and the frequency of events was a Swiss mathematician named Jacob Bernoulli, shortly before his death in 1705. What Bernoulli showed was that if the probability of getting heads is 50%, then the probability that the percentage of heads you actually got would *differ* from 50% by any given amount got smaller and smaller the more times you flipped the coin. You were more likely to get 50% heads if you flipped the coin 100 times than if you flipped it just twice. There's something fishy about this answer, though, since it uses ideas from probability to say what probabilities mean. If this seems confusing, it turns out you can do a little better. Bernoulli didn't realize this (in fact, it wasn't fully worked out until the twentieth century), but it is possible to prove that if the chance of getting heads when you flip a coin is 50%, and you flip a coin an *infinite* number of times, then it is (essentially) certain that half of the times will be heads. Or, for de Méré's strategy, if he played his dice game an infinite number of times, betting on 6 in every game, he would be essentially guaranteed to win 51.7477% of the games. This result is known as the law of large numbers. It underwrites one of the most important interpretations of probability.

Pascal was never much of a gambler himself, and so it is ironic that one of his principal mathematical contributions was in this arena. More ironic still is that one of the things he's most famous for is a bet that bears his name. At the end of 1654, Pascal had a mystical experience that changed his life. He stopped working on mathematics and devoted himself entirely to Jansenism, a controversial Christian movement prominent in France in the seventeenth century. He began to write extensively on theological matters. Pascal's Wager, as it is now called, first appeared in a note among his religious writings. He argued that you could think of the choice of whether to believe in God as a kind of gamble: either the Christian God exists or he doesn't, and a person's beliefs amount to a bet one way or the other. But before taking any bet, you want to know what the odds are and what happens if you win versus what happens if you lose. As Pascal reasoned, if you bet that God exists and you live your life accordingly, and you're right, you spend eternity in paradise. If you're wrong, you just die and nothing happens. So, too, if you bet against God and you win. But if you bet against God and you lose, you are damned to perdition. When he thought about it this way, Pascal decided the decision was an easy one. The downside of atheism was just too scary.

Despite his fascination with chance, Louis Bachelier never had much luck in life. His work included seminal contributions to physics, finance, and mathematics, and yet he never made it past the fringes of academic respectability. Every time a bit of good fortune came his way it would slip from his fingers at the last moment. Born in 1870 in Le Havre, a bustling port town in the northwest of France, young Louis was a promising student. He excelled at mathematics in *lycée* (basically, high school) and then earned his *baccalauréat ès sciences* — the equivalent of A-levels in Britain or a modern-day AP curriculum in the United States — in October 1888. He had a strong enough record that he could likely have attended one of France's selective *grandes écoles*, the French Ivy League, elite universities that served as prerequisites for life as a civil servant or intellectual. He came from a middle-class merchant family, populated by amateur scholars and artists. Attending a *grande école* would have opened intellectual and professional doors for Bachelier that had not been available to his parents or grandparents.

But before Bachelier could even apply, both of his parents died. He was left with an unmarried older sister and a three-year-old brother to care for. For two years, Bachelier ran the family wine business, until he was drafted into military service in 1891. It was not until he was released from the military, a year later, that Bachelier was able to return to his studies. By the time he returned to academia, now in his early twenties and with no family back home to support him, his options were limited. Too old to attend a *grande école,* he enrolled at the University of Paris, a far less prestigious choice.

Still, some of the most brilliant minds in Paris served as faculty at the university — it was one of the few universities in France where faculty could devote themselves to research, rather than teaching — and it was certainly possible to earn a first-rate education in the halls of the Sorbonne. Bachelier quickly distinguished himself among his peers. His marks were not the best at the university, but the small handful of students who bested him, classmates like Paul Langevin and Alfred-Marie Liénard, are now at least as famous as Bachelier himself, among mathematicians anyway. It was good company to be in. After finishing his undergraduate degree, Bachelier stayed at the University of Paris for his doctorate. His work attracted the attention of the best minds of the day, and he began to work on a dissertation — the one Samuelson later discovered, on speculation in financial markets — with Henri Poincaré, perhaps the most famous mathematician and physicist in France at the time.

Poincaré was an ideal person to mentor Bachelier. He had made substantial contributions to every field he had come in contact with, including pure mathematics, astronomy, physics, and engineering. Although he did attend a *grande école* as an undergraduate, like Bachelier he had done his graduate work at the University of Paris. He also had experience working outside of academia, as a mine inspector. Indeed, for most of his life he continued to work as a professional mining engineer, ultimately becoming the chief engineer of the French Corps de Mines, and so he was able to fully appreciate the importance of working on applied mathematics, even in areas so unusual (for the time) as finance. It would have been virtually impossible for Bachelier to produce his dissertation without a supervisor who was as wide-ranging and ecumenical as Poincaré. And more, Poincaré's enormous success had made him a cultural and political figure in France, someone who could serve as a highly influential advocate for a student whose research was difficult to situate in the then-current academic world.

And so it was that Bachelier wrote his thesis, finishing in 1900. The basic idea was that probability theory, the area of mathematics invented by Cardano, Pascal, and Fermat in the sixteenth and seventeenth centuries, could be used to understand financial markets. In other words, one could imagine a market as an enormous game of chance. Of course, it is now commonplace to compare stock markets to casinos, but this is only testament to the power of Bachelier's idea.

By any intellectual standard, Bachelier's thesis was an enormous success — and it seems that, despite what happened next, Bachelier knew as much. Professionally, however, it was a disaster. The problem was the audience. Bachelier was at the leading edge of a coming revolution — after all, he had just invented mathematical finance — with the sad consequence that none of his contemporaries were in a position to properly appreciate what he had done. Instead of a community of like-minded scholars, Bachelier was evaluated by mathematicians and mathematically oriented physicists. In later times, even these groups might have been sympathetic to Bachelier's project. But in 1900, Continental mathematics was deeply inward-looking. The general perception among mathematicians was that mathematics was just emerging from a crisis that had begun to take shape around 1860. During this period many well-known theorems were shown to contain errors, which led mathematicians to fret that the foundation of their discipline was crumbling. At issue, in particular, was the question of whether suitably rigorous methods could be identified, so as to be sure that the new results flooding academic journals were not themselves as flawed as the old. This rampant search for rigor and formality had poisoned the mathematical well so that applied mathematics, even mathematical physics, was looked at askance by mainstream mathematicians. The idea of bringing mathematics into a new field, and worse, of using intuitions from finance to drive the development of new mathematics, was abhorrent and terrifying.

Poincaré's influence was enough to shepherd Bachelier through his thesis defense, but even he was forced to conclude that Bachelier's essay fell too far from the mainstream of French mathematics to be awarded the highest distinction. Bachelier's dissertation received a grade of *honorable,* and not the better *très honorable.* The committee's report, written by Poincaré, reflected Poincaré's deep appreciation of Bachelier's work, both for the new mathematics and for its deep insights into the workings of financial markets. But it was impossible to grant the highest grade to a mathematics dissertation that, by the standards of the day, was not on a topic in mathematics. And without a grade of *très honorable* on his dissertation, Bachelier's prospects as a professional mathematician vanished. With Poincaré's continued support, Bachelier remained in Paris. He received a handful of small grants from the University of Paris and from independent foundations to pay for his modest lifestyle. Beginning in 1909, he was permitted to lecture at the University of Paris, but without drawing a salary.

The cruelest reversal of all came in 1914. Early that year, the Council of the University of Paris authorized the dean of the Faculty of Science to create a permanent position for Bachelier. At long last, the career he had always dreamed of was within reach. But before the position could be finalized, fate threw Bachelier back down. In August of that year, Germany marched through Belgium and invaded France. In response, France mobilized for war. On the ninth of September, the forty-four-year-old mathematician who had revolutionized finance without anyone noticing was drafted into the French army.

Imagine the sun shining through a window in a dusty attic. If you focus your eyes in the right way, you can see minute dust particles dancing in the column of light. They seem suspended in the air. If you watch carefully, you can see them occasionally twitching and changing directions, drifting upward as often as down. If you were able to look closely enough, with a microscope, say, you would be able to see that the particles were constantly jittering. This seemingly random motion, according to the Roman poet Titus Lucretius (writing in about

60 B.C.), shows that there must be tiny, invisible particles—he called them "primordial bits"—buffeting the specks of dust from all directions and pushing them first in one direction and then another.

Two thousand years later, Albert Einstein made a similar argument in favor of the existence of atoms. Only he did Lucretius one better: he developed a mathematical framework that allowed him to precisely describe the trajectories a particle would take if its twitches and jitters were really caused by collisions with still-smaller particles. Over the course of the next six years, French physicist Jean-Baptiste Perrin developed an experimental method to track particles suspended in a fluid with enough precision to show that they indeed followed paths of the sort Einstein predicted. These experiments were enough to persuade the remaining skeptics that atoms did indeed exist. Lucretius's contribution, meanwhile, went largely unappreciated.

The kind of paths that Einstein was interested in are examples of Brownian motion, named after Scottish botanist Robert Brown, who noted the random movement of pollen grains suspended in water in 1826. The mathematical treatment of Brownian motion is often called a random walk—or sometimes, more evocatively, a drunkard's walk. Imagine a man coming out of a bar in Cancun, an open bottle of sunscreen dribbling from his back pocket. He walks forward for a few steps, and then there's a good chance that he will stumble in one direction or another. He steadies himself, takes another step, and then stumbles once again. The direction in which the man stumbles is basically random, at least insofar as it has nothing to do with his purported destination. If the man stumbles often enough, the path traced by the sunscreen dripping on the ground as he weaves his way back to his hotel (or just as likely in another direction entirely) will look like the path of a dust particle floating in the sunlight.

In the physics and chemistry communities, Einstein gets all the credit for explaining Brownian motion mathematically, because it was his 1905 paper that caught Perrin's eye. But in fact, Einstein was five years too late. Bachelier had already described the mathematics of random walks in 1900, in his dissertation. Unlike Einstein, Bachelier had little interest in the random motion of dust particles as they bumped into atoms. Bachelier was interested in the random movements of stock prices.

Imagine that the drunkard from Cancun is now back at his hotel. He gets out of the elevator and is faced with a long hallway, stretching off to both his left and his right. At one end of the hallway is room 700; at the other end is room 799. He is somewhere in the middle, but he has no idea which way to go to get to his room. He stumbles to and fro, half the time moving one way down the hall, and half the time moving in the opposite direction. Here's the question that the mathematical theory of random walks allows you to answer: Suppose that with each step the drunkard takes, there is a 50% chance that that step will take him a little farther toward room 700, at one end of the long hallway, and a 50% chance that it will take him a little farther toward room 799, at the other end. What is the probability that, after one hundred steps, say, or a thousand steps, he is standing in front of a given room?

To see how this kind of mathematics can be helpful in understanding financial markets, you just have to see that a stock price is a lot like our man in Cancun. At any instant, there is a chance that the price will go up, and a chance that the price will go down. These two possibilities are directly analogous to the drunkard stumbling toward room 700, or toward room 799, working his way up or down the hallway. And so, the question that mathematics can answer in this case is the following: If the stock begins at a certain price, and it undergoes a random walk, what is the probability that the price will be a particular value after some fixed period of time? In other words, which door will the price have stumbled to after one hundred, or one thousand, ticks?

This is the question Bachelier answered in his thesis. He showed that if a stock price undergoes a random walk, the probability of its taking any given value after a certain period of time is given by a curve known as a normal distribution, or a bell curve. As its name suggests, this curve looks like a bell, rounded at the top and widening at the bottom. The tallest part of this curve is centered at the starting price, which means that the most likely scenario is that the price will be somewhere near where it began. Farther out from this center peak, the curve drops off quickly, indicating that large changes in price are less likely. As the stock price takes more steps on the random walk, however, the curve progressively widens and becomes less tall overall, indicating that over time, the chances that the stock will vary from its initial value increase. A picture is priceless here, so look at Figure 1 to see how this works.

Thinking of stock movements in terms of random walks is astound-

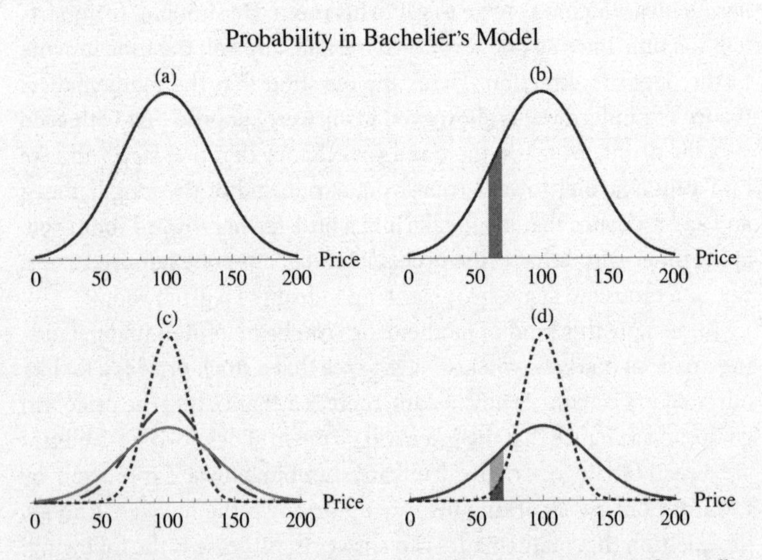

Figure 1: Bachelier discovered that if the price of a stock undergoes a random walk, the probability that the price will take a particular value in the future can be calculated from a curve known as a normal distribution. These plots show how that works for a stock whose price is $100 now. Plot (a) is an example of a normal distribution, calculated for a particular time in the future, say, five years from now. The probability that, in five years, the price of the stock will be somewhere in a given range is given by the area underneath the curve—so, for instance, the area of the shaded region in plot (b) corresponds to the probability that the stock will be worth somewhere between $60 and $70 in five years. The shape of the plot depends on how long into the future you are thinking about projecting. In plot (c), the dotted line would be the plot for a year from now, the dashed line for three years, and the solid line for five years from now. You'll notice that the plots get shorter and fatter over time. This means that the probability that the stock will have a price very far from its initial price of $100 gets larger, as can be seen in plot (d). Notice that the area of the shaded region under the solid line, corresponding to the probability that the price of the stock will be between $60 and $70 five years from now, is much larger than the area of the shaded region below the dotted line, which corresponds to just one year from now.

ingly modern, and it seems Bachelier was essentially unprecedented in conceiving of the market in this way. And yet on some level, the idea seems crazy (perhaps explaining why no one else entertained it). Sure, you might say, I believe the mathematics. If stock prices move randomly, then the theory of random walks is well and good. But why would you ever assume that markets move randomly? Prices go up on good news; they go down on bad news. There's nothing random about it. Bachelier's basic assumption, that the likelihood of the price ticking up at a given instant is always equal to the likelihood of its ticking down, is pure bunk.

This thought was not lost on Bachelier. As someone intimately familiar with the workings of the Paris exchange, Bachelier knew just how strong an effect information could have on the prices of securities. And looking backward from any instant in time, it is easy to point to good news or bad news and use it to explain how the market moves. But Bachelier was interested in understanding the probabilities of *future* prices, where you don't know what the news is going to be. Some future news might be predictable based on things that are already known. After all, gamblers are very good at setting odds on things like sports events and political elections — these can be thought of as predictions of the likelihoods of various outcomes to these chancy events. But how does this predictability factor into market behavior? Bachelier reasoned that any predictable events would already be reflected in the current price of a stock or bond. In other words, if you had reason to think that something would happen in the future that would ultimately make a share of Microsoft worth more — say, that Microsoft would invent a new kind of computer, or would win a major lawsuit — you should be willing to pay more for that Microsoft stock now than someone who didn't think good things would happen to Microsoft, since you have reason to expect the stock to go up. Information that makes positive future events seem likely pushes prices up *now;* information that makes negative future events seem likely pushes prices down *now.*

But if this reasoning is right, Bachelier argued, then stock prices *must* be random. Think of what happens when a trade is executed at a given price. This is where the rubber hits the road for a market. A trade means that two people — a buyer and a seller — were able to agree on a price. Both buyer and seller have looked at the available information and have decided how much they think the stock is worth to them, but with an important caveat: the buyer, at least according to Bachelier's logic, is buying the stock at that price because he or she thinks that in the future the price is likely to go up. The seller, meanwhile, is selling at that price because he or she thinks the price is more likely to go down. Taking this argument one step further, if you have a market consisting of many informed investors who are constantly agreeing on the prices at which trades should occur, the current price of a stock can be interpreted as the price that takes into account all possible information. It is the price at which there are just as many informed people willing to bet that the price will go up as are willing to bet that the price will go down. In other words, at any moment, the current price is the price at which all available information suggests that the probability of the stock ticking up and the probability of the stock ticking down are both 50%. If markets work the way Bachelier argued they must, then the random walk hypothesis isn't crazy at all. It's a necessary part of what makes markets run.

This way of looking at markets is now known as the efficient market hypothesis. The basic idea is that market prices always reflect the true value of the thing being traded, because they incorporate all available information. Bachelier was the first to suggest it, but, as was true of many of his deepest insights into financial markets, few of his readers noted its importance. The efficient market hypothesis was later rediscovered, to great fanfare, by University of Chicago economist Eugene Fama, in 1965. Nowadays, of course, the hypothesis is highly controversial. Some economists, particularly members of the so-called Chicago School, cling to it as an essential and irrefutable truth. But you don't have to think too hard to realize it's a little fishy. For instance, one consequence of the hypothesis is that there can't be any speculative bubbles, because a bubble can occur only if the market price for something becomes unmoored from the thing's actual value. Anyone who remembers the dot-com boom and bust in the late nineties/early 2000s, or anyone who has tried to sell a house since about 2006, knows that prices don't behave as rationally as the Chicago School would have us believe. Indeed, most of the day-to-day traders I've spoken with find the idea laughable.

But even if markets aren't always efficient, as they surely aren't, and even if sometimes prices get quite far out of whack with the values of the goods being traded, as they surely do, the efficient market hypothesis offers a foothold for anyone trying to figure out how markets work. It's an assumption, an idealization. A good analogy is high school physics, which often takes place in a world with no friction and no gravity. Of course, there's no such world. But a few simplifying assumptions can go a long way toward making an otherwise intractable problem solvable — and once you solve the simplified problem, you can begin to ask how much damage your simplifying assumptions do. If you want to understand what happens when two hockey pucks bump into each other on an ice rink, assuming there's no friction won't get you into too much trouble. On the other hand, assuming there's no friction when you fall off a bicycle could lead to some nasty scrapes. The situation is the same when you try to model financial markets: Bachelier *begins* by assuming something like the efficient market hypothesis, and he makes amazing headway. The next step, which Bachelier left to later generations of people trying to understand finance, is to figure out when the assumption of market efficiency fails, and to come up with new ways to understand the market when it does.

It seems that Samuelson was the only recipient of Savage's postcards who ever bothered to look Bachelier up. But Samuelson was impressed enough, and influential enough, to spread what he found. Bachelier's writings on speculation became required reading among Samuelson's students at MIT, who, in turn, took Bachelier to the far corners of the world. Bachelier was officially canonized in 1964, when Paul Cootner, a colleague of Samuelson's at MIT, included an English translation of Bachelier's thesis as the first essay in an edited volume called *The Random Character of Stock Market Prices*. By the time Cootner's collection was published, the random walk hypothesis had been ventured independently and improved upon by a number of people, but Cootner was unambiguous in assigning full credit for the idea to Bachelier. In Cootner's words, "So outstanding is [Bachelier's] work that we can say that the study of speculative prices has its moment of glory at its moment of conception."

In many ways, Samuelson was the ideal person to discover Bachelier and to effectively spread his ideas. Samuelson proved to be one of the most influential economists of the twentieth century. He won the second Nobel Prize in economics, in 1970, for "raising the level of analysis in economic science," the prize committee's code for "turning economics into a mathematical discipline." Indeed, although he studied economics both as an undergraduate at the University of Chicago and as a graduate student at Harvard, he was deeply influenced by a mathematical physicist and statistician named E. B. Wilson. Samuelson met Wilson while still a graduate student. At the time, Wilson was a professor of "vital statistics" at the Harvard School of Public Health, but he had spent the first twenty years of his career as a physicist and engineer at MIT. Wilson had been the last student of J. W. Gibbs, the first great American mathematical physicist — indeed, the first recipient of an American PhD in engineering, in 1863 from Yale. Gibbs is most famous for having helped lay the foundations of thermodynamics and statistical mechanics, which attempt to explain the behavior of ordinary objects like tubs of water and car engines in terms of their microscopic parts.

Through Wilson, Samuelson became a disciple of the Gibbsian tradition. His dissertation, which he wrote in 1940, was an attempt to rewrite economics in the language of mathematics, borrowing extensively from Gibbs's ideas about statistical thermodynamics. One of the central aims of thermodynamics is to offer a description of how the behavior of particles, the small constituents of ordinary matter, can be aggregated to describe larger-scale objects. A major part of this analysis is identifying variables like temperature or pressure that don't make sense with regard to individual particles but can nonetheless be used to characterize their collective behavior. Samuelson pointed out that economics can be thought of in essentially the same way: an economy is built out of people going around making ordinary economic decisions. The trick to understanding large-scale economics — macroeconomics — is to try to identify variables that characterize the economy as a whole — the inflation rate, for instance — and then work out the relationship of these variables to the individuals who make up the economy. In 1947, Samuelson published a book based on his dissertation at Harvard, called *Foundations of Economic Analysis.*

Samuelson's book was groundbreaking in a way that Bachelier's thesis never could have been. When Bachelier was studying, economics was only barely a professional discipline. In the nineteenth century, it was basically a subfield of political philosophy. Numbers played little role until the 1880s, and even then they entered only because some philosophers became interested in measuring the world's economies to better compare them. When Bachelier wrote his thesis, there was essentially no field of economics to revolutionize — and of the few economists there were, virtually none would have been able to understand and appreciate the mathematics Bachelier used.

Over the next forty years, economics matured as a science. Early attempts to measure economic quantities gave way to more sophisticated tools for relating different economic quantities to one another — in part because of the work of Irving Fisher, the first American economist and another student of Gibbs's at Yale. For the first decades of the twentieth century, research in economics was sporadic, with some mild support from European governments during World War I, as the needs of war pushed governments to try to enact policies that would increase production. But the discipline fully came into its own only during the early 1930s, with the onset of the Depression. Political leaders across Europe and the United States came to believe that something had gone terribly wrong with the world's economy and sought expert advice on how to fix it. Suddenly, funding for research spiked, leading to a large number of university and government positions. Samuelson arrived at Harvard on the crest of this new wave of interest, and when his book was published, there was a large community of researchers who were at least partially equipped to understand its significance. Samuelson's book and a subsequent textbook, which has since gone on to become the best-selling economics book of all time, helped others to appreciate what Bachelier had accomplished nearly half a century earlier.

In modern parlance, what Bachelier provided in his thesis was a *model* for how market prices change with time, what we would now call the random walk model. The term *model* made its way into economics during the 1930s, with the work of another physicist turned economist, Jan Tinbergen. (Samuelson was the second Nobelist in economics; Tinbergen was the first.) The term was already being used in physics, to refer to something just shy of a full physical theory. A theory, at least as it is usually thought of in physics, is an attempt to completely and accurately describe some feature of the world. A model, meanwhile, is a kind of simplified picture of how a physical process or system works. This was more or less how Tinbergen used the term in economics, too, although his models were designed specifically to devise ways of predicting relationships between economic variables, such as the relationship between interest rates and inflation or between different wages at a single firm and the overall productivity of that firm. (Tinbergen famously argued that a company would become less productive if the income of the highest-paid employee was more than five times the income of the lowest-paid employee — a rule of thumb largely forgotten today.) Unlike in physics, where one often works with full-blown theories, mathematical economics deals almost exclusively with models.

By the time the Cootner book was published in 1964, the idea that market prices follow a random walk was well entrenched, and many economists recognized that Bachelier was responsible for it. But the random walk model wasn't the punch line of Bachelier's thesis. He thought of it as preliminary work in the service of his real goal, which was developing a model for pricing options. An option is a kind of derivative that gives the person who owns the option the right to buy (or sometimes sell) a specific security, such as a stock or bond, at a predetermined price (called the strike price), at some future time (the expiration date). When you buy an option, you don't buy the underlying stock directly. You buy the right to trade that stock at some point in the future, but at a price that you agree to in the present. So the price of an option should correspond to the value of the right to buy something at some time in the future.

Even in 1900, it was obvious to anyone interested in trading that the value of an option had to have something to do with the value of the underlying security, and it also had to have something to do with the strike price. If a share of Google is trading at $100, and I have a contract that entitles me to buy a share of Google for $50, that option is worth at least $50 to me, since I can buy the share of Google at the discounted rate and then immediately sell it at a profit. Conversely, if the option gives me the right to buy a share at $150, the option isn't going to do me much good — unless, of course, Google's stock price shoots up to above $150. But figuring out the precise relationship was a mystery. What should the right to do something in the future be worth now?

Bachelier's answer was built on the idea of a fair bet. A bet is considered fair, in probability theory, if the average outcome for both people involved in the bet is zero. This means that, on average, over many repeated bets, both players should break even. An unfair bet, meanwhile, is when one player is expected to lose money in the long run. Bachelier argued that an option is itself a kind of bet. The person selling the option is betting that between the time the option is sold and the time it expires, the price of the underlying security will fall beneath the strike price. If that happens, the seller wins the bet — that is, makes a profit on the option. The option buyer, meanwhile, is betting that at some point the price of the underlying security will exceed the strike price, in which case *the buyer* makes a profit, by exercising the option and immediately selling the underlying security. So how much should an option cost? Bachelier reasoned that a fair price for an option would be the price that would make it a fair bet.

In general, to figure out whether a bet is fair, you need to know the probability of every given outcome, and you need to know how much you would gain (or lose) if that outcome occurred. How much you gain or lose is easy to work out, since it's just the difference between the strike price on the option and the market price for the underlying security. But with the random walk model in hand, Bachelier also knew how to calculate the probabilities that a given stock would exceed (or fail to exceed) the strike price in a given time window. Putting these two elements together, Bachelier showed just how to calculate the fair price of an option. Problem solved.

There's an important point to emphasize here. One often hears that markets are unpredictable because they are random. There is a sense in which this is right, and Bachelier knew it. Bachelier's random walk model indicates that you can't predict whether a given stock is going to go up or down, or whether your portfolio will profit. But there's another sense in which some features of markets are predictable precisely *because* they are random. It's because markets are random that you can use Bachelier's model to make probabilistic predictions, which, because of the law of large numbers — the mathematical result that Bernoulli discovered, linking probabilities with frequency — give you information about how markets will behave in the long run. This kind of prediction is useless for someone speculating on markets directly, because it doesn't let the speculator pick which stocks will be the winners and which the losers. But that doesn't mean that statistical predictions can't help investors — just consider Bachelier's options pricing model, where the assumption that markets for the underlying assets are random is the key to its effectiveness.

That said, even a formula for pricing options isn't a guaranteed trip to the bank. You still need a way to use the information that the formula provides to guide investment decisions and gain an edge on the market. Bachelier offered no clear insight into how to incorporate his options pricing model in a trading strategy. This was one reason why Bachelier's options pricing model got less attention than his random walk model, even after his thesis was rediscovered by economists. A second reason was that options remained relatively exotic for a long time after he wrote his dissertation, so that even when economists in the fifties and sixties became interested in the random walk model, the options pricing model seemed quaint and irrelevant. In the United States, for instance, most options trading was illegal for much of the twentieth century. This would change in the late 1960s and again in the early 1970s. In the hands of others, Bachelier-style options pricing schemes would lay the foundations of fortunes.

Bachelier survived World War I. He was released from the military on the last day of 1918. On his return to Paris, he discovered that his position at the University of Paris had been eliminated. But overall, things were better for Bachelier after the war. Many promising young mathematicians had perished in battle, opening up university positions. Bachelier spent the first years after the war, from 1919 until 1927, as a visiting professor, first in Besançon, then in Dijon, and finally in Rennes. None of these were particularly prestigious universities, but they offered him paid teaching positions, which were extremely rare in France. Finally, in 1927, Bachelier was appointed to a full professorship at Besançon, where he taught until he retired in 1937. He lived for nine years more, revising and republishing work that he had written earlier in his career. But he stopped doing original work. Between the time he became a professor and when he died, Bachelier published only one new paper.

An event that occurred toward the end of Bachelier's career, in 1926 (the year before he finally earned his permanent position), cast a pall over his final years as a teacher and may explain why he stopped publishing. That year, Bachelier applied for a permanent position at Dijon, where he had been teaching for several years. One of his colleagues, in reviewing his work, became confused by Bachelier's notation. Believing he had found an error, he sent the document to Paul Lévy, a younger but more famous French probability theorist. Lévy, examining only the page on which the error purportedly appeared, confirmed the Dijon mathematician's suspicions. Bachelier was blacklisted from Dijon. Later, he learned of Lévy's part in the fiasco and became enraged. He circulated a letter claiming that Lévy had intentionally blocked his career without understanding his work. Bachelier earned his position at Besançon a year later, but the damage had been done and questions concerning the legitimacy of much of Bachelier's work remained. Ironically, in 1941, Lévy read Bachelier's final paper. The topic was Brownian motion, which Lévy was also working on. Lévy found the paper excellent. He corresponded with Bachelier, returned to Bachelier's earlier work, and discovered that he, not Bachelier, had been wrong about the original point — Bachelier's notation and informal style had made the paper difficult to follow, but it was essentially correct. Lévy wrote to Bachelier and they reconciled, probably sometime in 1942.

Bachelier's work is referenced by a number of important mathematicians working in probability theory during the early twentieth century. But as the exchange with Lévy shows, many of the most influential people working in France during Bachelier's lifetime, including people who worked on topics quite close to Bachelier's specialties, were either unaware of him or dismissed his work as unimportant or flawed. Given the importance that ideas like his have today, one is left to conclude that Bachelier was simply too far ahead of his time. Soon after his death, though, his ideas reappeared in the work of Samuelson and his students, but also in the work of others who, like Bachelier, had come to economics from other fields, such as the mathematician Benoît Mandelbrot and the astrophysicist M.F.M. Osborne. Change was afoot in both the academic and financial worlds that would bring these later prophets the kind of recognition that Bachelier never enjoyed while he was alive.

CHAPTER 2

Swimming Upstream

MAURY OSBORNE'S MOTHER, Amy Osborne, was an avid gardener. She was also a practical woman. Rather than buy commercial fertilizer, she would go out to the horse pastures near her home, in Norfolk, Virginia, to collect manure and bring it back for her garden. And she didn't approve of idleness. Whenever she caught one of her sons lazing about, she was quick to assign a job: paint the porch, cut the grass, dig a hole to mix up the soil. When Osborne was young, he liked the jobs. Painting and hole-digging were fun enough, and other jobs, like cutting the grass, were unpleasant but better than sitting around doing nothing. Whenever he got bored, he would go to his mother and ask what he could do, and she would give him a job.

One day, she pointed out that the ice truck had just passed. The truck was pulled by a horse, which meant that there would be nice big piles of manure on the road. "So you go and collect that horse manure and mix it up with the hose to make liquid manure and pour it on my chrysanthemums," she told him. Osborne didn't much like this assignment. It was the middle of the day and all of his friends were out and about, and when they saw him they yelled out and teased him. Red-faced and fuming, he dutifully collected the manure in a big bucket, then went back to his house. He pulled out the hose, filled the bucket with water, and began to liquefy the manure. It was a gross, smelly job, and Osborne was feeling irritated and embarrassed at having to do it in the first place. Then all of a sudden, as he was stirring, the liquefied manure splashed out of the bucket and soaked him. It was a major turning point: there, covered in fresh horse manure, Osborne decided that he would never ask anyone what to do again — he would figure out what *he* wanted to do and do *that*.

As far as his scientific career went, Osborne kept his pledge. He was initially trained as an astronomer, calculating things like the orbits of planets and comets. But he never felt constrained by academic boundaries. Shortly before the United States entered World War II, Osborne left graduate school to work at the Naval Research Lab (NRL) on problems related to underwater sound and explosions. The work had very little to do with astronomical observation, but Osborne thought it would be interesting. Indeed, before the war was over, he took up several different projects. In 1944, for example, he wrote a paper on the aerodynamics of insect wings. In the 1940s, entomologists had no idea why insects could fly. Their bodies seemed to be too heavy for the amount of lift generated by flapping wings. Well, Osborne had some time on his hands, and so, instead of asking the navy what he should do, he decided he'd spend his time solving the problem of insect flight. And he succeeded: he showed, for the first time, that if you took into account both the lift produced by insect wings and the drag on the wings, you could come up with a pretty good explanation for why insects can fly and how they control their motion.

After World War II, Osborne went further still. He approached the head of the NRL's Sound Division, where he still worked, and told him that anyone working for the government could get their work done in two hours a day. Bold words for one's boss, you might think. But Osborne pressed further. He said that even two hours of work a day was more than he wanted to do for the government. He had a problem of his own that he wanted to work on. Osborne made it clear that this new project had nothing at all to do with naval interests, but he said he wanted to work on it anyway. And amazingly, his boss said, "Go right ahead."

Osborne remained at the NRL for nearly thirty more years, but from that conversation on, he worked exclusively on his own projects. In most cases, these projects had little or no direct bearing on the navy, and yet the NRL continued to support him throughout his career. The work ran the gamut from foundational problems in general relativity and quantum mechanics to studies of deep ocean currents. But his most influential work, the work for which he is best known today, was on another topic entirely. In 1959, Osborne published a paper entitled "Brownian Motion in the Stock Market." Though Bachelier had written on this very subject sixty years earlier, his work was still essentially unknown to physicists or financiers (aside from a few people in Samuelson's circle). To readers of Osborne's paper, the suggestion that physics had something to say about finance was entirely novel. And it wasn't long before people in academia and on Wall Street began to take notice.

Any way you look at it, Bachelier's work was genius. As a physicist, he anticipated some of Einstein's most influential early work — work that would later be used to definitively prove the existence of atoms and usher in a new era in science and technology. As a mathematician, he developed probability theory and the theory of random processes to such a high level that it would take three decades for other mathematicians to catch up. And as a mathematical analyst of financial markets, Bachelier was simply without peer. It is exceptionally rare in any field for someone to present so mature a theory with so little precedent. In a just world, Bachelier would be to finance what Newton is to physics. But Bachelier's life was a shambles, in large part because academia couldn't countenance so original a thinker.

Just a few short decades later, though, Maury Osborne was thriving in a government-sponsored lab. He could work on anything he liked, in whatever style he liked, without facing any of the institutional resistance that plagued Bachelier throughout his career. Bachelier and Osborne had much in common: both were incredibly creative; both had the originality to find questions that hadn't occurred to previous researchers and the technical skills to make them tractable. But when Osborne happened on the same problem that Bachelier had addressed in his thesis — the problem of predicting stock prices — and proceeded to work out a remarkably similar solution, he did so in a completely different environment. "Brownian Motion in the Stock Market" was an unusual article. But in the United States in 1959, it was acceptable, even *encouraged*, for a physicist of Osborne's station to work on such problems. As Osborne put it, "Physicists essentially could do no wrong." Why had things changed?

Nylon. American women were first introduced to nylon at the 1939 New York World's Fair, and they were smitten. A year later, on May 15, 1940, when nylon stockings went on sale in New York, 780,000 pairs were sold on the first day, and 40 million pairs by the end of the week. At year's end, Du Pont, the company that invented and manufactured nylon, had sold 64 million pairs of nylon stockings in the United States alone. Nylon was strong and lightweight. It tended to shed dirt and it was water resistant, unlike silk, which was the preferred material for hosiery before nylon hit the scene. Plus, it was much cheaper than either silk or wool. As the *Philadelphia Record* put it, nylon was "more revolutionary than [a] martian attack."

But nylon had revolutionary consequences far beyond women's fashion or fetishists' lounges. The initiative at Du Pont that led to the invention of nylon — along with a handful of other research programs begun in the 1930s by companies such as Southern California Edison, General Electric, and Sperry Gyroscope Company, and universities such as Stanford and Berkeley — quietly ushered in a new research culture in the United States.

In the mid-1920s, Du Pont was a decentralized organization, with a handful of largely independent departments, each of which had its own large research division. There was also a small central research unit, essentially a holdover from an earlier period in Du Pont's history, headed by a man named Charles Stine. Stine faced a problem. With so many large, focused research groups at the company, each performing whatever services its respective department required, the need for an additional research body was shaky at best. If the central research unit was going to survive, never mind grow, Stine needed to articulate a mission for it that would justify its existence. The solution he finally came upon and implemented in 1927 was the creation of an elite, fundamental research team within the central research unit. The idea was that many of Du Pont's industrial departments relied on a core of basic science. But the research teams in these departments were too focused on the immediate needs of their businesses to engage in fundamental research. Stine's team would work on these orphaned scientific challenges over the long term, laying the foundation for future applied, industrial work. Stine landed a chemist from Harvard, named Wallace Carothers, to head this new initiative.

Carothers and a team of young PhDs spent the next three years exploring and exhaustively documenting the properties of various polymers—chemical compounds composed of many small, identical building blocks (called monomers) strung together like a chain. During these early years, the work proceeded unfettered by commercial considerations. The central research unit at Du Pont functioned as a pure, academic research laboratory. But then, in 1930, Carothers's team had two major breakthroughs. First, they discovered neoprene, a synthetic rubber. Later that same month, they discovered the world's first fully synthetic fiber. Suddenly Stine's fundamental research team had the potential to make real money for the company, fast. Du Pont's leadership took notice. Stine was promoted to the executive committee and a new man, Elmer Bolton, was put in charge of the unit. Bolton had previously headed research in the organic chemistry department and, in contrast to Stine, he had much less patience for research without clear applications. He quickly moved research on neoprene to his old department, which had considerable experience in rubber, and encouraged Carothers's team to focus on synthetic fibers. The initial fiber turned out to have some poor properties: it melted at low temperatures and dissolved in water. But by 1934, under pressure from his new boss, Carothers came up with a new idea for a polymer that he thought would be stable when spun into a fiber. Five weeks later, one of his lab assistants produced the first nylon.

Over the next five years, Du Pont embarked on a crash program to scale up production and commercialize the new fiber. Nylon began life as an invention in a pure research lab (even though, under Bolton's direction, Carothers was looking for such fibers). As such, it represented cutting-edge technology, based on the most advanced chemistry of the time. But it was not long before it was transformed into a commercially viable, industrially produced product. This process was essentially new: as much as nylon represented a major breakthrough in polymer chemistry, Du Pont's commercialization program was an equally important innovation in the industrialization of basic research. A few important features distinguished the process. First, it required close collaboration among the academic scientists in the central research unit, the industrial scientists in the various departments' research divisions, and the chemical engineers responsible for building a new plant and actually producing the nylon. As the different teams came together to solve one problem after another, the traditional boundaries between basic and applied research, and between research and engineering, broke down.

Second, Du Pont developed all of the stages of manufacturing of the polymer in parallel. That is, instead of waiting until the team fully understood the first stage of the process (say, the chemical reaction by which the polymer was actually produced) and only then moving on to the next step (say, developing a method for spinning the polymer into a fiber), teams worked on all of these problems at once, each team taking the others' work as a "black box" that would produce a fixed output by some not-yet-known method. Working in this way further encouraged collaboration between different kinds of scientists and engineers because there was no way to distinguish an initial basic research stage from later implementation and application stages. All of these occurred at once. Finally, Du Pont began by focusing on a single product: women's hosiery. Other uses of the new fiber, including lingerie and carpets, to name a few, were put off until later. This deepened everyone's focus, at every level of the organization. By 1939, Du Pont was ready to reveal the product; by 1940, the company could produce enough of it to sell.

The story of nylon shows how the scientific atmosphere at Du Pont changed, first gradually and then rapidly as the 1930s came to a close, to one in which pure and applied work were closely aligned and both were valued. But how did this affect Osborne, who didn't work at Du Pont? By the time nylon reached shelves in the United States, Europe was already engaged in a growing war effort—and the U.S. government was beginning to realize that it might not be able to remain neutral. In 1939, Einstein wrote a letter to Roosevelt warning that the Germans were likely to develop a nuclear weapon, prompting Roosevelt to launch a research initiative, in collaboration with the United Kingdom, on the military uses of uranium.

After the Japanese attack on Pearl Harbor, on December 7, 1941, and Germany's declaration of war on the United States four days later, work on nuclear weapons research accelerated rapidly. Work on uranium continued, but in the meantime, a group of physicists working at Berkeley had isolated a new element—plutonium—that could also be used in nuclear weapons and that could, at least in principle, be mass produced more easily than uranium. Early in 1942, Nobel laureate Arthur Compton secretly convened a group of physicists at the University of Chicago, working under the cover of the "Metallurgical Laboratory" (Met Lab), to study this new element and to determine how to incorporate it into a nuclear bomb.

By August 1942, the Met Lab had produced a few milligrams of plutonium. The next month, the Manhattan Project began in earnest: General Leslie Groves of the Army Corps of Engineers was assigned command of the nuclear weapons project; Groves promptly made Berkeley physicist J. Robert Oppenheimer, who had been a central part of the Met Lab's most important calculations, head of the effort. The Manhattan Project was the single largest scientific endeavor ever embarked on: at its height, it employed 130,000 people, and it cost a total of $2 billion (about $22 billion in today's dollars). The country's entire physics community rapidly mobilized for war, with research departments at most major universities taking part in some way, and with many physicists relocating to the new secret research facility at Los Alamos.

Groves had a lot on his plate. But one of the very biggest problems involved scaling up production of plutonium from the few milligrams the Met Lab had produced to a level sufficient for the mass production of bombs. It is difficult to overstate the magnitude of this challenge. Ultimately, sixty thousand people, nearly half of the total staff working on the Manhattan Project, would be devoted to plutonium production. When Groves took over in September 1942, the Stone and Webster Engineering Corporation had already been contracted to build a large-scale plutonium enrichment plant in Hanford, Washington, but Compton, who still ran the Met Lab, didn't think Stone and Webster was up to the task. Compton voiced his concern, and Groves agreed that Stone and Webster didn't have the right kind of experience for the job. But then, where *could* you find a company capable of taking a few milligrams of a brand-new, cutting-edge material and building a production facility that could churn out tons of the stuff, fast?

At the end of September 1942, Groves asked Du Pont to join the project, advising Stone and Webster. Two weeks later, Du Pont agreed to do much more: it took full responsibility for the design, construction, and operation of the Hanford plant. The proposed strategy? Do for plutonium precisely what Du Pont had done for nylon. From the beginning, Elmer Bolton, who had led the just-finished nylon project as head of the central research unit, and several of his closest associates took leadership roles in the plutonium project. And just like nylon, the industrialization of plutonium was an enormous success: in a little over two years, the nylon team ramped up production of plutonium a million-fold.

Implementing the nylon strategy was not a simple task, nor was it perfectly smooth. To produce plutonium on a large scale, you need a full nuclear reactor, which, in 1942, had never been built (though plans were in the works). This meant that, even more than with nylon, new technology and basic science were essential to the development of the Hanford site, which in turn meant that the physicists at the Met Lab felt they had a stake in the project and took Du Pont's role to be "just" engineering. They believed that as nuclear scientists, they were working at the very pinnacle of human knowledge. As far as they were concerned, industrial scientists and engineers were lesser beings. Needless to say, they did not take well to the new chain of command.

The central problem was that the physicists significantly underestimated the role engineers would have to play in constructing the site. They argued that Du Pont was putting up unnecessary barriers to research by focusing on process and organization. Ironically, this problem was solved by giving the physicists *more* power over engineering: Compton negotiated with Du Pont to let the Chicago physicists review and sign off on the Du Pont engineers' blueprints. But once the physicists saw the sheer scale of the project and began to understand just how complex the engineering was going to be, many gained an appreciation of the engineers' role — and some even got interested in the more difficult problems.

Soon, scientists and engineers were engaged in an active collaboration. And just as the culture at Du Pont had shifted during the nylon project — as the previously firm boundaries between science and engineering began to crumble — the collaboration between physicists and engineers at the Hanford site quickly broke down old disciplinary barriers. In building the plutonium facility, Du Pont effectively exported its research culture to an influential group of theoretical and experimental physicists whose pre- and postwar jobs were at universities, not in industry. And the shift in culture survived. After the war, physicists were accustomed to a different relationship between pure and applied work. It became perfectly acceptable for even top theoretical physicists to work on real-world problems. And equally important, for basic research to be "interesting," physicists needed to sell their colleagues on its possible applications.

Du Pont's nylon project wasn't the only place where a new research culture developed during the 1930s, and the Hanford site and Met Lab weren't the only government labs at which physicists and engineers were brought into close contact during World War II. Similar changes took place, for similar reasons, at Los Alamos, the Naval Research Lab, the radiation labs at Berkeley and MIT, and in many other places around the country as the needs of industry, and then the military, forced a change in outlook among physicists. By the end of the war, the field had been transformed. No longer could the gentleman-scientist of the late nineteenth or early twentieth century labor under the illusion that his work was above worldly considerations. Physics was now too big and too expensive. The wall between pure physics and applied physics had been demolished.

Born in 1916, Osborne was exceptionally precocious. He finished high school at fifteen, but his parents wouldn't let him attend college so young, so he spent a year in prep school — which he hated — before going on to the University of Virginia to major in astrophysics. The intellectual independence and broad, innate curiosity that would later characterize his scientific career were apparent early on. After his first year of college, for instance, Osborne decided he'd had enough of studying. So one day that summer, after finishing a job at the McCormick Observatory in Charlottesville, Virginia, he decided to drop out of school. Instead of going back to UVA, he would spend some time doing physical labor. He told his parents his plan, and apparently they knew better than to try to talk him out of it, because they contacted a family friend with a farm in West Virginia and Osborne went there to work for the year. But he was sent home for Christmas, followed shortly by a note from the farm's owner saying that she had had quite enough of him. Osborne spent the rest of the year pushing a wheelbarrow around Norfolk, helping the director of physical education for the Norfolk school district regrade playgrounds. The year of hard labor convinced Osborne that academic life wasn't so bad after all. He returned to UVA the following September.

After college, Osborne headed west to Berkeley for a graduate program in astronomy. There he met and worked closely with luminaries in the physics department, including Oppenheimer. This is where Osborne was when war broke out in Europe in 1939. By the spring of 1941, many physicists, Oppenheimer included, were beginning to think about the war effort, including the possible use of nuclear weapons. Osborne saw the writing on the wall. Recognizing that he would likely be drafted, he attempted to enlist — but he was rejected because he wore thick glasses (early in the war effort, recruiters could afford to be picky). So he sent an application to the NRL, which offered him a job in its Sound Division. He packed his bags and headed home to

Virginia to work in a government lab at the moment the government was most prepared to support creative, interdisciplinary research.

Osborne began "Brownian Motion in the Stock Market" with a thought experiment. "Let us imagine a statistician," he wrote, "trained perhaps in astronomy and totally unfamiliar with finance, is handed a page of the *Wall Street Journal* containing the N.Y. Stock Exchange transactions for a given day." Osborne began thinking about the stock market around 1956, after his wife, Doris (also an astronomer), had given birth to a second set of twins — the Osbornes' eighth and ninth children, respectively. Osborne decided he had better start thinking about financing the future. One can easily imagine Osborne going down to the store and picking up a copy of the day's *Wall Street Journal.* He would have brought it home, sat down at the kitchen table, and opened it to the pages that reported the previous day's transactions. Here he would have found hundreds, perhaps thousands, of pieces of numerical data, in columns labeled with strange, undefined terms.

The statistician trained in astronomy wouldn't have known what the labels meant, or how to interpret the data, but that was fine. Numerical data didn't scare him. After all, he'd seen page after page of data recording the nightly motions of the heavens. The difficulty was figuring out how the numbers related to each other, determining which numbers gave information about which other numbers, and seeing if he could make any predictions. He would, in effect, be building a model from a set of experimental data, which he'd done dozens of other times. So Osborne would have adjusted his glasses, rolled up his sleeves, and dived right in. Lo and behold, he discovered some familiar patterns: the numbers corresponding to price behaved just like a collection of particles, moving randomly in a fluid. As far as Osborne could tell, these numbers could have come from dust exhibiting Brownian motion.

In many ways, Osborne's first, and most lasting, contribution to the theory of stock market behavior recapitulated Bachelier's thesis. But there was a big difference. Bachelier argued that from moment to moment stock prices were as likely to go up by a certain small amount as to go down by that same amount. From this he determined that stock prices would have a normal distribution. But Osborne dismissed this idea immediately. (Samuelson did, too — in fact, he called this aspect of Bachelier's work absurd.) A simple way to test the hypothesis that the probabilities governing future stock prices are determined by a normal distribution would be to select a random collection of stocks and plot their prices. If Bachelier's hypothesis were correct, one would expect the stock prices to form an approximate bell curve. But when Osborne tried this, he discovered that prices *don't* follow a normal distribution at all! In other words, if you looked at the data, Bachelier's findings were ruled out right away. (To his credit, Bachelier *did* examine empirical data, but a certain unusual feature of the market for *rentes* — specifically, that their prices changed very slowly, and never by very much — made his model seem more effective than it actually was.)

So what did Osborne's price distribution look like? It looked like a hump with a long tail on one side, but virtually no tail on the other side. This shape doesn't look much like a bell, but it was familiar enough to Osborne. It's what you get, not if prices themselves are normally distributed, but if *the rate of return* is normally distributed. The rate of return on a stock can be thought of as the average percentage by which the price changes each instant. Suppose you took $200, deposited $100 in a savings account, and used the other $100 to buy some stock. A year from now, you probably wouldn't have the $200 (you might have more or less), because of interest accrued in the savings account, and because of changes in the price of the stock that you purchased. The rate of return on the stock can be thought of as the interest rate that your bank would have had to pay (or charge) to keep the balances in your two accounts equal. It is a way of capturing the change in the price of a stock relative to its initial price.

The rate of return on a stock is related to the change in price by a mathematical operation known as a logarithm. For this reason, if rates of return are normally distributed, the probability distribution of stock prices should be given by something known as a log-normal distribution. (See Figure 2 for what this looks like.) The log-normal distribution was the funny-looking hump with a tail that Osborne found when he plotted actual stock prices. The upshot of this analysis was that it's the rate of return that undergoes a random walk, and not the price. This observation corrects an immediate, damning problem with Bachelier's model. If stock prices are normally distributed, with the width of the distribution determined by time, then Bachelier's model predicts that after a sufficiently long period of time, there would always be a chance that any given stock's price would become negative. But this is impossible: a stockholder cannot lose more than he or she initially invested. Osborne's model doesn't have this problem. No matter how negative the rate of return on a stock becomes, the price itself never becomes negative — it just gets closer and closer to zero.

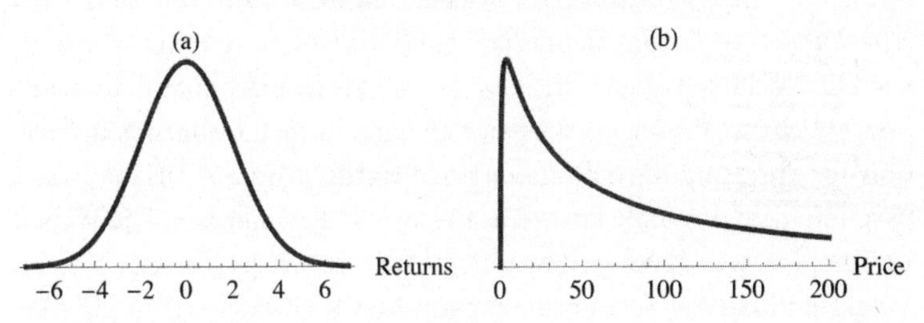

Figure 2: Osborne argued that rates of return, not prices, are normally distributed. Since price and rate of return are related by a logarithm, Osborne's model implies that prices should be *log-normally* distributed. These plots show what these two distributions look like at some time in the future, for a stock whose price is $10 now. Plot (a) is an example of a normal distribution over rates of return, and plot (b) is the associated log-normal distribution for the prices, given those probabilities for rates of return. Note that on this model, rates of return can be negative, but prices never are.

Osborne had another reason for believing that the rate of return, not the price itself, should undergo a random walk. He argued that investors don't really care about the absolute movement of stocks. Instead, they care about the percentage change. Imagine that you have a stock that is worth $10, and it goes up by $1. You've just made 10%. Now imagine the stock is worth $100. If it goes up by $1, you're happy — but not as happy, since you've made only 1%, even though you've made a dollar in both cases. If the stock starts at $100, it has to go all the way up to $110 for an investor to be as pleased as if the $10 stock went up to $11. And logarithms respect this relativized valuation: they have the nice property that the difference between log(10) and log(11) is equal to the difference between log(100) and log(110). In other words, the rate of return is the same for a stock that begins at $10 and goes up to $11 as for a stock that begins at $100 and goes up to $110. Statisticians would say that the logarithm of price has an "equal interval" property: the difference between the logarithms of two prices corresponds to the difference in psychological sensation of gain or loss corresponding to the two prices.

You might notice that the argument in the last paragraph, which is just the argument Osborne gave in "Brownian Motion in the Stock Market," has a slightly surprising feature: it says that we should be interested in the logarithms of prices because logarithms of prices *better reflect how investors feel about their gains and losses.* In other words, it's not the objective value of the change in a stock price that matters, it's how an investor reacts to the price change. In fact, Osborne's motivation for choosing logarithms of price as his primary variable was a psychological principle known as the Weber-Fechner law. The Weber-Fechner law was developed by nineteenth-century psychologists Ernst Weber and Gustav Fechner to explain how subjects react to different physical stimuli. In a series of experiments, Weber asked blindfolded men to hold weights. He would gradually add more weight to the weights the men were already holding, and the men were supposed to say when they felt an increase. It turned out that if a subject started out holding a small weight — just a few grams — he could tell when a few more grams were added. But if the subject started out with a larger weight, a few more grams *wouldn't* be noticed. It turned out that the smallest noticeable change was proportional to the starting weight. In other words, the psychological effect of a change in stimulus isn't determined by the absolute magnitude of the change, but rather by its change relative to the starting point.

So, as Osborne saw it, the fact that investors seem to care about percentage change rather than absolute change reflected a general psychological fact. More recently, people have criticized mathematical modeling of financial markets using methods from physics on the grounds that the stock market is composed of people, not quarks or pulleys. Physics is fine for billiard balls and inclined planes, even for space travel and nuclear reactors, but as Newton said, it cannot predict the madness of men. This kind of criticism draws heavily on ideas from a field known as behavioral economics, which attempts to understand economics by drawing on psychology and sociology. From this point of view, markets are all about the foibles of human beings — they cannot be reduced to the formulas of physics and mathematics. For this reason alone, Osborne's argument is historically interesting, and I think telling. It shows that mathematical modeling of financial markets is not only consistent with thinking about markets in terms of the psychology of investors, but that the best mathematical models will be ones that, like Osborne's and unlike Bachelier's, take psychology into account. Of course, Osborne's psychology was primitive, even by the standards of 1959. (The Weber-Fechner law was already a century old when Osborne applied it, and much subsequent research had been conducted on how human subjects register change.) Modern economics can draw on far more sophisticated theories of psychology than the Weber-Fechner law, and later in the book we will see some examples where it has. But bringing in new insights from psychology and related fields only strengthens our ability to use mathematics to reliably model financial markets, by guiding us to make more realistic assumptions and by helping us identify situations where the current crop of models might be expected to fail.

Osborne was accustomed to working with the very finest physicists of his day, and he could not be cowed by authority. If he worked out the solution to a problem, or if he believed he understood something, he argued his case forcefully. In early 1946, for instance, Osborne became interested in relativity theory. To learn as much about the theory as he could, he picked up a book by Einstein, *The Meaning of Relativity*, in which Einstein offered an argument about how much dark matter could exist in the universe. Dark matter — literally, stuff in the universe that doesn't seem to emit or reflect light, which means that we can't see it directly — was first discovered in the 1930s, by its effects on the rotation of galaxies. Devotees of popular physics know that today, dark matter is one of the most puzzling mysteries in all of cosmology. Observations of other galaxies suggest that the vast majority of the matter in the universe is unobservable, something that is not explained by any of our best physical theories.

Einstein proposed a simple way of figuring out the lower bound for the total amount of dark matter in the universe. He argued that the density of dark matter in the universe as a whole was at least as much as the density within a galaxy (or rather, a group of galaxies known as a cluster). Osborne decided he didn't buy the argument. For one, Einstein seemed to be making a series of bad assumptions. Worse still, the best evidence that anyone had in 1946 showed that most dark matter was restricted to certain parts of a galaxy, with basically no dark matter in empty space (this still seems to be true). So if anything, you should expect the density of dark matter to be higher in a galaxy than in space as a whole.

By 1946, most people, if they disagreed with an argument of Einstein's pertaining to relativity and astrophysics, would assume they had misunderstood something. Einstein was already a cultural icon. But Osborne took no heed of such things. When he understood something, he understood it, and no amount of reputation or authority could intimidate him. So Osborne wrote Einstein a letter in which he very politely suggested that Einstein's argument didn't make any sense. Einstein replied by restating his argument from the book. So Osborne wrote again. Einstein conceded that his argument was problematic but thought the conclusion remained sound, and so he offered another argument. Once again, Osborne refuted it. At the end of a half-dozen-letter correspondence, it was clear that Einstein was unconvinced by Osborne. But it was equally clear to Osborne that Einstein's argument in the book failed, and that he didn't have any other good arguments up his sleeve.*

Osborne approached his work in economics in the same spirit. Unconcerned about his lack of background in economics or finance,

* I think most physicists today, if they read the letters, would say that Osborne got the better of the exchange.

Osborne presented his research with an engineer's confidence. He published "Brownian Motion in the Stock Market" in a journal called *Operations Research.* It was not an economics journal, but enough economists and economically minded mathematicians read it that Osborne's research quickly garnered attention. Some of this was positive, but it was not unambiguously so. Indeed, when Osborne published his first paper on finance, he was unaware of Bachelier or Samuelson, or any of a handful of economists who had, in one way or another, anticipated the idea that stock prices are random. Many economists pointed out his lack of originality — so many that Osborne was forced to publish a second paper just a few months after the first, in which he presented a brief history of the idea that markets are random, giving full credit to Bachelier for coming up with the idea first, but also defending his own formulation.

Osborne stood his ground, and rightfully so. Despite connections with earlier work, his papers on randomness in the stock market were sufficiently original that Samuelson later gave him credit for developing the modern version of the random walk hypothesis at the same time that Samuelson and his students were working on it. More importantly still, Osborne approached his model as a true empirical scientist, trained to handle data. He developed and applied a series of statistical tests designed to corroborate his version of the Brownian motion model. Other researchers, such as the statistician Maurice Kendall, who in 1953 showed that stock prices were as likely to go up as to go down, had done empirical work on the randomness of stock prices. But Osborne was the first to demonstrate the importance of the log-normal distribution to markets. He was also the first to clearly articulate a model for how stock market randomness worked and how it could be used to derive probabilities for future prices (and rates of return), all while providing convincing data that this particular model of the markets captured how markets really behave. And despite the early reservations about Osborne's originality, economists soon recognized that he brought theory and evidence together in a way that simply hadn't been done before. When Paul Cootner at MIT collected the most important papers on the random walk hypothesis for his 1964 volume — the volume that contained the first English translation of

Bachelier's thesis — he included two papers by Osborne. One was the 1959 paper on Brownian motion; the other was a paper that expanded on and generalized the earlier work.

By the time Osborne began thinking about markets, he had published fifteen papers in physics and related topics. He had held a permanent position at the NRL for a decade and a half and had rubbed shoulders with some of the best physicists of the mid-twentieth century, as both colleague and correspondent. And yet, Osborne still didn't have a PhD, in physics or in anything else. He had left grad school in 1941 to join the NRL without finishing his degree. On one level, a doctorate didn't mean much for a person like Osborne; he had a fulfilling career in physics even without a doctorate, and no one seemed to doubt his credentials as a researcher. His work spoke for itself. He decided, however, during the mid-fifties, that he wanted to finish his degree, at least in part because it would guarantee him a promotion at the NRL. And so Osborne followed many of his colleagues at the NRL to the physics department at the University of Maryland. There he could finish his graduate work without giving up his position at the lab.

Osborne's first attempt at a dissertation was on a topic in astronomy. (Usually graduate students write a dissertation proposal. Osborne ignored this step. He wrote entire dissertations.) He brought the dissertation to the physics department head, who promptly rejected it because too many people were interested in the topic and Osborne's research wasn't original enough. So Osborne wrote a second dissertation, based on his research on the stock market. The department head rejected this, too, on the grounds that it wasn't physics. As Osborne would later put it, "You are supposed to do original research, but if you get too original, they don't know what's going on." Stock market research may have been acceptable work for a physicist within the government research community, where applied work of any stripe was highly valued. But it still wasn't "physics" from the perspective of a traditional academic department. And so, though Osborne was received more favorably by the scientific community than Bachelier, he was still something of a maverick for working on financial modeling.

Even after having two dissertations rejected, Osborne wasn't ready to give up. He sent "Brownian Motion in the Stock Market" off to *Operations Research* and set to writing a third dissertation. For this project, he returned to a problem he had been working on just before he began to think about the stock market. The third idea concerned the migratory efficiency of salmon. Salmon spend most of their lives in the ocean. But when it comes time to breed, they return to their birthplaces, often up to a thousand miles upstream of the ocean, to spawn and die. But after leaving the ocean, they no longer eat. Osborne realized that this meant that one could figure out how efficiently a salmon can swim by looking at the distances traveled and the fat lost on arrival. The idea was to think of a salmon as a boat that was traveling a certain distance without refueling.

When he finished this third dissertation and submitted it, he again received a lukewarm reaction. It was not clear that this third dissertation was any more "physics" than the second one had been. Ultimately, however, the dissertation was accepted. The university was in the process of applying for a large grant in biophysics (the study of the physics of biological systems), and the administration wanted to have evidence of expertise in that field. And so, in 1959, almost twenty years after he had first moved to the NRL and the same year that "Brownian Motion in the Stock Market" appeared in print, Osborne finally received a doctorate (and a much-deserved promotion at the NRL).

The work on migratory salmon bears a surprising connection to Osborne's work on financial markets. His model of how salmon swim upriver included analysis at several different time scales. There were effects corresponding to how well the salmon were able to swim over short distances, which depended on things like the strength of the current in the river at a given moment. There were also effects that you couldn't see clearly just by looking at a salmon swimming for a few feet or yards but became apparent when you looked at a salmon traveling over, say, a thousand miles. The first kind of effect might be called "fast" fluctuations in the salmon's efficiency; the second might be called "slow" fluctuations. The trouble was that the data were much better on the slow fluctuations. It's easy to record how many salmon, roughly, have reached a given point at a given time; it is much harder to record just how well any given salmon is making headway as a river's current changes.

Osborne had worked out a theoretical model that tried to explain both the slow and fast fluctuations, and to show how they related to each other. And he wanted to figure out a way to test the model. Getting better data on individual salmon would have been one way to do this — but it would have been difficult, and Osborne didn't have any idea where to start. A second possibility was to find another system that might show both the fast and slow fluctuations that Osborne wanted to study, to see if the same model described that system as well. This second option seemed much more appealing, but Osborne needed an appropriate system. When he sat down to figure out how to understand the stock quotes in the *Wall Street Journal*, he soon realized that markets, too, have different scales of fluctuations. Some market forces, like the details of how an exchange works or the interactions of traders, can affect how prices change over the course of a day. These are like the fast fluctuations that salmon experience from one river bend to the next. But there are other forces affecting markets, things like business cycles and government interest rates, that become apparent only when you step back and look at a longer time period. These are slow fluctuations. It turned out the financial world was the perfect place to look for data that could be used to test Osborne's ideas about how these different kinds of fluctuations affect one another.

The process worked in the other direction, too. After developing the migratory salmon model in the context of stock market prices, and after tweaking the model to better fit the data he had used to test it, he applied it to a problem in physics. Osborne proposed a new model for deep ocean currents. Specifically, he was able to explain how the random motion of water molecules (fast fluctuations in the language of the salmon paper) could give rise to variations in apparently systematic large-scale phenomena, like currents (slow fluctuations). For Osborne, work in physics and finance were intrinsically linked.

It is tempting to overstate both the reception of Osborne's work and his direct influence, because as we shall see, his ideas would ultimately revolutionize financial markets. Still, his work did not make the splash on Wall Street that more developed versions of his ideas would, in the hands of other researchers just a short time later. Osborne was a transitional figure. He was read widely by academics and some theoretically minded practitioners, but Wall Street was not yet ready to move firmly in the direction that Osborne's work suggested. In part the difficulty was that Osborne believed that his model of market randomness implied that it was impossible to predict how individual stock prices would change with time; unlike Bachelier, Osborne didn't connect his work to options, where understanding the statistical properties of markets can help you identify when options are correctly priced. Indeed, reading "Brownian Motion in the Stock Market" and Osborne's later work, one gets the sense that there is *no* way to profit from the stock market. Prices are unpredictable. The speculator's average gain is zero. Investing is a losing proposition.

Later, people would look at Osborne's work and see something more optimistic. If you know that stock prices are essentially random, then, as Bachelier pointed out, you can figure out the value of options or other derivatives based on those stocks. Osborne didn't take his work in this direction — at least, not until the late 1970s, when others had already made similar moves. Instead, he spent much of the rest of his career trying to figure out the ways in which stock prices *aren't* random. In other words, after tying himself to the enormously controversial claim that stock prices represent "unrelieved bedlam" (his words, in many of his articles), Osborne systematically and exhaustively searched for order and predictability.

He had some limited success. He showed that the volume of trading — the number of trades that take place in any given stretch of time — isn't constant, as one would naively assume in a Brownian motion model. Instead, there are peaks in volume at the beginning and end of a trading day, over the course of an average trading week, and over the course of a month. (All of these variations, incidentally, represent just the kind of "slow fluctuations" Osborne had explored with his migratory salmon — applied not to prices, but to numbers of trades.) These variations arise from what Osborne took to be another principle of market psychology, that investors have limited attention spans. They get interested in a stock, they make a lot of trades and send the volume of trades way up, and then they gradually stop paying attention and volume decreases. If you allow for variations in volume, you have to change the underlying assumptions of the random walk model, and you get a new, more accurate model of how stock prices evolve, which Osborne called the "extended Brownian motion" model.

In the mid-sixties, Osborne and a collaborator showed that at any instant, the chances that a stock will go up are not necessarily the same as the chances that the stock will go down. This assumption, you'll recall, was an essential part of the Brownian motion model, where a step in one direction is assumed to be just as likely as a step in the other. Osborne showed that if a stock went up a little bit, its next motion was much more likely to be a move back down than another move up. Likewise, if a stock went *down*, it was much more likely to go up in value in its next change. That is, from moment to moment the market is much more likely to *reverse* itself than to continue on a trend. But there was another side to this coin. If a stock moved in the same direction *twice*, it was much more likely to continue in that direction than if it had moved in a given direction only once. Osborne argued that the infrastructure of the trading floor was responsible for this kind of nonrandomness, and Osborne went on to suggest a model for how prices change that took this kind of behavior into account.

This was a hallmark of Osborne's work, and it was one of the reasons he's such an important figure in the story of physics and finance. The idea that prices are equally likely to move up or down was part of Osborne's version of the efficient market hypothesis, a central assumption of his original model. When he realized this assumption didn't hold, he began to look for ways to tweak the model to account for a more realistic assumption, based on what he had learned about real markets. Osborne was explicit from the beginning that this was his methodology, in keeping with the kinds of theoretical work he was familiar with in astronomy and fluid dynamics. In those fields, most problems are much too hard to solve all at once. Instead, you begin by studying the data and then make simplifying assumptions to derive simple models. But this is only the first step. Next, you check carefully to find places where your simplifying assumptions break down and try to figure out, again by focusing on the data, how these failures of your assumptions produce problems for the model's predictions.

When Osborne described his original Brownian motion model, he specifically indicated what assumptions he was making. He pointed out that if the assumptions were no good, there was no guarantee that the model would be, either. What Osborne and other physicists understood was that a model isn't "flawed" when the assumptions underlying it fail. But it does mean you have more work to do. Once you've proposed a model, the next step is to figure out when the assumptions fail and how badly. And if you discover that the assumptions fail regularly, or under specific circumstances, you try to understand the ways in which they fail and the reasons for the failures. (For instance, Osborne showed that price changes aren't independent. This is especially true during market crashes, when a series of downward ticks makes it very likely that prices will continue to fall. When this kind of herding effect is present, even Osborne's extended Brownian motion model is going to be an unreliable guide.) The model-building process involves constantly updating your best models and theories in light of new evidence, pulling yourself up by the bootstraps as you progressively understand whatever you're studying—be it cells, hurricanes, or stock prices.

Not everyone who has worked with mathematical models in finance has been as sensitive to the importance of this methodology as Osborne was, which is one of the principal reasons why mathematical models have sometimes been associated with financial ruin. If you continue to trade based on a model whose assumptions have ceased to be met by the market, and you lose money, it is hardly a failure of the model. It's like attaching a car engine to a plane and being disappointed when it doesn't fly.

Despite the patterns in stock prices that Osborne was able to discover, he remained convinced that in general, there was no reliable way to make profitable forecasts about future market behavior. There was, however, one exception. Ironically, it had nothing to do with the sophisticated models that he developed during the 1960s. Instead, his optimism was based on a way of reading the mind of the markets, by studying the behavior of traders.

Osborne noticed that a great preponderance of ordinary investors placed their orders at whole-number prices — $10, or $11 say. But stocks were valued in units of 1/8 of a dollar. This meant that a trader could look at his book and see that there were a lot of people who wanted to buy a stock at, say, $10. He could then buy it at $10 1/8, knowing that at the end of the day the stock wouldn't drop below $10 because there were so many people willing to buy at that threshold. So at worst, the trader would lose $1/8; at best, the stock would go up, and he could make a lot. Conversely, he could see that a lot of people wanted to sell at, say, $11, and so he could sell at $10 7/8 with confidence that the most he could lose would be $1/8 if the stock went up instead of down. This meant that if you went through a day's trades and looked for trades at $1/8 above or below whole-dollar amounts, you could gather which stocks the experts thought were "hot" because so many other people were interested.

It turned out that what the experts thought was hot was a great indicator of how stocks would do — a much better indicator than anything else Osborne had studied. Based on these observations, Osborne proposed the first trading program of a sort that could be plugged into a computer to run on its own. But in 1966, when he came up with the idea, no one was using computers to make decisions. It would take decades for Osborne's idea and others like it to be tested in the real world.

CHAPTER 3

From Coastlines to Cotton Prices

SZOLEM MANDELBROJT WAS THE very model of a modern mathematician. An expert in analysis (the area of abstract mathematics that includes, among other things, standard college calculus), he had studied in Paris with the best of the best, including Emile Picard and Henri Lebesgue. He was a founding member of a group of French mathematicians who, under the pseudonym Nicolas Bourbaki, endeavored to bring the highest possible level of rigor and abstraction to the field; the group's collected works set the tone for two generations of mathematicians. When his mentor, Jacques Hadamard, one of the most famous mathematicians of the late nineteenth century, retired from his position at the prestigious Collège de France, the Collège invited Mandelbrojt to replace him. He was a serious man, doing serious work.

Or at least he would have been doing serious work if his nephew hadn't been constantly hounding him. In 1950, Benoît Mandelbrot was a doctoral student at the University of Paris, Szolem's alma mater, seeking (Szolem imagined) to follow in his eminent uncle's footsteps. When Szolem first learned that Benoît wanted to pursue mathematics, he was thrilled. But gradually, Szolem began to question Benoît's seriousness. Despite his uncle's advice, Benoît showed no interest in the pressing mathematical matters of the day. His work lacked the rigor that had brought Szolem such success. Worst of all, Benoît seemed intent on *geometrical* methods, which every self-respecting mathematician knew had been abandoned a century before because they had led so many people astray. Real mathematics couldn't be done by drawing pictures.

Benoît's father, Szolem's oldest brother, had helped raise Szolem. He had supported Szolem through graduate school, creating opportunities Szolem would never have had otherwise. To Szolem, then, Benoît was more like a brother than a nephew, and Szolem felt that he owed Benoît his continued patience and support. But Szolem was at the end of his rope. Benoît just wasn't getting it. He had as much mathematical aptitude as anyone, but when it came to picking projects, he was hopeless.

One day, while Benoît was in his office talking about his crazy dissertation ideas, Szolem snapped. He reached into his trash can and pulled out a discarded paper. If Benoît wanted to work on trash, Szolem had plenty of it to give him — a whole bin filled with papers of no interest or importance. "This is for you," he said dismissively. "That's the kind of silly stuff you like."

Szolem must have hoped his dramatic gesture would knock some sense into his young nephew. But the plan backfired magnificently. Benoît took the paper — a review of a recent book by a Harvard linguist named George Kingsley Zipf — and studied it carefully on his way home. Zipf was a famously eccentric character and few took him seriously. He had spent his career arguing for a universal law of physical, social, and linguistic phenomena. Zipf's law said that if you constructed a list of all the things in some natural category, say, all of the cities in France, or all of the libraries in the world, and ranked them according to their size — you might rank cities by population; libraries, by collection size — you would always find that the size of each thing on the list was related to its rank on the list. In particular, the second thing on each list would always be about half the size of the first thing, the third thing on the list would be about a third the size of the first thing, and so on. The review that Benoît read focused on a particular example of the law in action: Zipf had gone through and counted how often various words appeared in various texts. He then showed that if you ordered the words by how often they appeared in a piece of writing, you usually found that the most common word appeared about twice as often as the second most common word, three times as often as the third most common word, and so on for all of the words in the document.

Szolem was right that Zipf's work was just the kind of thing his nephew would be interested in. But he was wrong that it was trash — or at least that it was all trash. Zipf's law is a peculiar combination of estimation and numerology and Zipf was a crank. But there was a gem hidden in his book: Zipf had worked out a formula that could be used to calculate how often a particular word would appear in a book, given its rank on the list and the total number of different words appearing in the text. Mandelbrot quickly realized that the formula could be improved upon, and moreover that it had some unexpected and interesting mathematical properties. Despite the resistance of the brightest lights in the mathematical establishment, his uncle included, Mandelbrot wrote a dissertation on Zipf's law and its applications. He did so without an advisor and received his degree only by pushing his thesis through the university's bureaucratic channels himself. It was highly irregular.

Indeed, Mandelbrot made a career out of the highly irregular, both in his impetuous rejection of the mathematical community and in his topics of study. Whereas the vast majority of mathematicians focus on shapes that are "smooth," the kinds of shapes you can make out of Play-Doh, Mandelbrot's most famous discovery, which he named "fractal geometry," arose out of the study of jagged and fractured shapes, like the surface of a mountain or a shard of broken glass. This work on fractal shapes made Mandelbrot realize that there are varieties of randomness in nature that are far more extreme than the kind of randomness you get by flipping a coin over and over again — with consequences for virtually all mathematical science, including finance.

Mandelbrot was a revolutionary. Even today, decades after his most important papers, his ideas remain radical, with mainstream scientists in many fields still debating them. The situation is particularly striking in economics, where Mandelbrot's central ideas have gone down like a bitter pill. If they are correct, almost everything traditional economists believe about markets is fundamentally flawed. It didn't help that Mandelbrot was uncompromising, both as a person and as a scientist, never bending to academic pressures. He often found himself at the fringes of respectability: esteemed, though never as highly as he deserved; criticized and dismissed as much for his style as for the unconventionality of his work. Yet over the past four decades, as Wall Street and the scientific community have encountered new, seemingly insurmountable challenges, Mandelbrot's insights into randomness have seemed ever more prescient — and more essential to understand.

Benoît Mandelbrot was born in 1924, to Lithuanian parents living in Warsaw, Poland. Although his father was a businessman, two of his uncles (including Szolem) were scholars. Many of his father's other relatives were, in Mandelbrot's words, "wise men" with no particular employment, but with a group of followers in the community who would trade money or goods in exchange for advice or learning. His mother, meanwhile, was also well educated, trained as a physician. As a boy, Mandelbrot often felt that he was expected to pursue an academic life of one sort or another, though his father urged him to choose a practical form of scholarship, such as engineering or applied science.

Despite the family's focus on learning, however, the young Mandelbrot had a very unusual education. His parents' first child, a daughter, died very young when an epidemic ripped through Warsaw. Benoît's mother developed a deep fear of childhood illnesses and sought to protect her two young sons from her daughter's fate. So rather than send Benoît to school, she hired one of his uncles to tutor him. This uncle, though related by marriage, was cast firmly in the mold of Mandelbrot's father's family: well educated and unemployed, with esoteric interests. He despised rote learning, so didn't bother to teach Benoît such mundane topics as arithmetic or the alphabet (indeed, in a speech he gave after receiving the Wolf Prize for physics, Mandelbrot admitted that he still had trouble multiplying, as he had never learned his multiplication tables). Instead, the uncle encouraged creative thought and voracious reading. Mandelbrot spent most of his time playing chess and studying maps.

Warsaw was hit hard by the Depression—worse than western Europe or the United States—and Mandelbrot's father's clothing business was essentially destroyed in 1931. His father then moved to France, hoping that the slightly better economic situation there would enable him to support his wife and sons from afar. With their large extended family in Warsaw, however, the Mandelbrots were strongly tied to the city. The hope was that Benoît's father would eventually be able to move back to Poland and reestablish his business there. But as the 1930s droned on and the Depression worsened, Poland became increasingly unsettled. Ethnic and political violence grew. As Jews, the Mandelbrots realized that Warsaw had become dangerous for them. Benoît's mother packed what belongings she could and followed her husband to Paris. Though a difficult decision at the time, the move to Paris almost certainly bought the Mandelbrots their lives: of the more than 3 million Jews who lived in Poland before World War II, only a few hundred thousand survived the Holocaust.

Szolem was already in Paris when Benoît's father arrived. He had moved to France in 1919, a refugee of an entirely different sort. In the immediate aftermath of World War I, mathematics in Poland was dominated by a brilliant young mathematician named Wacław Sierpiński. Sierpiński worked on a topic known as set theory. He was militant about his preferred style of mathematics and powerful enough to dictate the terms of success for any graduate student in Warsaw. Later in life, Szolem may have seemed unbearably rigid to the geometrically minded Mandelbrot, but Sierpiński was too formal even for Szolem. Refusing to work on the topics Sierpiński required, Szolem fled to Paris, where the prevailing mathematical ideology was more in line with his own. Ironically, Sierpiński was also the discoverer of an unusual geometrical object known as the Sierpiński triangle—an early example of a fractal.

It wasn't until Mandelbrot arrived in Paris that he had the opportunity to interact with his famous mathematician uncle. Mandelbrot was eleven years old. Though the two would later have their differences, their early relationship was deeply formative. Since Mandelbrot spoke little French, he was placed two grades behind his age level. To keep him interested in his education and to encourage his talents, Szolem fed him bits of mathematics. It was largely Szolem's influence during this period that pushed Mandelbrot toward mathematics. Despite the difficult economic and political situation, under Szolem's tutelage Benoît found a way to thrive in his new home.

Unfortunately, it would not last. In 1940, Germany invaded France. And once again, the Mandelbrots were forced to flee.

How long is Britain's coastline? This might seem like a simple question — one that could be easily settled, say, by a team of competent surveyors. As it turns out, however, the question is more complicated than it appears. There's a deep puzzle built into it, sometimes known as the coastline paradox. To figure out the length of a coastline, you need to take some measurements, presumably with some sort of ruler. The puzzle concerns how *long* your ruler needs to be. Suppose you started with a single enormous ruler that stretched from Cape Wrath, at the northernmost tip of Scotland, all the way down to Penzance, at the southwestern tip of Cornwall. This would give you an estimate of the length of the coastline.

But not a very good one. A coastline is hardly a straight line. The coast of Britain dips in at the Bristol Channel and the Irish Sea, jutting out again near Wales, so taking one very long ruler isn't going to give an accurate measurement. To get a better measurement, you would want to use a somewhat smaller ruler — one that could easily accommodate the additional length that the various peninsulas and bays add to the coast. You might try adding up the distances from, say, Penzance to Bristol, and then from Bristol to St. David's in Wales, and then from St. David's to Carmel Head at the northwestern tip of Wales, and so on all the way up the coast. This total distance would be a lot longer than the first distance you calculated, but it would be more accurate.

Now, though, a pattern begins to emerge. This smaller ruler, it turns out, underestimates the length in the same way the original long ruler did. Using the smaller ruler, you miss Cardigan Bay altogether, not to mention the dozens of smaller harbors and inlets along the Cornish and Welsh coasts. To account for *these* features, which turn out to add rather a lot of distance, you need a smaller ruler still. But again, the same problem arises. In fact, no matter what size ruler you pick, the answer you get by measuring the coastline with that ruler is *always* too small. In other words, you can always get a larger answer to the question by picking a smaller ruler.

This is where the paradox arises. It is often the case that choosing more precise instruments gives you a better measurement of something. You can get a sense of how hot a pot of water is by sticking your finger in it. An alcohol thermometer would do the job even better, and a high-tech digital thermometer would bring the accuracy to within a fraction of a degree. There is a sense in which the imprecise tools are adding measurement *error,* and as you devise better and better instruments, you home in on the real temperature. But with a coastline, no matter how precise your measuring device — that is, no matter how small your ruler — your measurement is always much too small. In some sense, a coastline doesn't *have* a length, or at least not in the way that simple shapes like a line or a circle do.

Mandelbrot addressed the coastline paradox in a groundbreaking paper in 1967. It was one of his first attempts to describe a fractal shape — as, indeed, a coastline turns out to be, though Mandelbrot didn't coin the term until 1975. Coastlines (and other fractals) are remarkable from a mathematical point of view because they have a property called self-similarity. To say that something is self-similar is to say that it is composed of pieces that look just like the whole; these pieces in turn are composed of still smaller pieces that *also* look like the whole, and so on ad infinitum. If you begin with the whole west coast of Britain and carve it up into several pieces, you will notice that each of these *also* looks like a coastline; just like the full coastline, the smaller stretches of coast have their own little inlets and peninsulas. And if you break up one of these smaller bits of coast further, the still-smaller pieces exhibit all the same features of the larger structures.

Once you start looking for self-similarity, you quickly realize it's a ubiquitous feature of nature. A mountaintop looks much like a whole mountain in miniature; a tree branch looks like a little tree, with smaller branches of its own; river systems are built out of smaller rivers and estuaries. The principle even seems to extend to the social world. As Mandelbrot later pointed out, a battle is made up of smaller skirmishes, and a war is composed of battles, each a microcosm of the war as a whole.

When World War II broke out, the Mandelbrots fled Paris, where they expected the fighting to be quite intense, and settled in a town called Tulle, in the region of France known as Corrèze. Once again, the Mandelbrots showed great foresight, not to mention luck: they left Paris in late 1939, mere months before the Nazi invasion of France. Tulle turned out to be an extremely fortuitous choice. It was far enough south that it would soon become part of unoccupied (Vichy) France.

The Vichy government cooperated with the Germans, but anti-Semitism in the south was less virulent than in the German-occupied territories. For a few years, at least, Mandelbrot was able to attend high school in Tulle. He was now fluent in French and he quickly moved through school, catching up to his peers by the time the Germans took control in 1942. Still, the Mandelbrots lived in constant fear of deportation. In 1940, the Vichy government had begun to review the status of all immigrants naturalized after 1927. They stripped about fifteen thousand (mostly Jews) of their citizenship, as a precursor to sending them to German concentration camps. Though the Mandelbrots managed to escape notice in little Tulle, the threat was ever present.

Matters became worse in 1942. On November 8, the British and American armies invaded French North Africa. In response, the Germans occupied southern France, anticipating an assault on continental Europe. With the German army came the Gestapo, and as southern France became a staging ground for defensive Panzer divisions, even Tulle became a minor battleground. Though it was home to only a few thousand people, Tulle was the traditional capital of the region. As the German presence in southern France increased, Tulle became a place of strategic interest to both the vestiges of the Vichy government and the leaders of the resistance. The Mandelbrots could no longer rely on the obscurity of their little town for safety.

In his autobiographical writings and interviews, Mandelbrot often spoke of the war's impact on his education. After finishing secondary school in 1942, he found himself unable to proceed to a *grande école* because his movements were so constrained. (Here his education is reminiscent of Bachelier's, who was also unable to attend a *grande école.*) But Mandelbrot never went into detail about his experiences during this period, except to say that the year and a half after finishing school was "very, very rough" and he had "several very close calls with disaster."

Since further schooling was out of the question, and because he needed to maintain a low profile, he avoided cities and moved often. He lived with members of the resistance, who took him in and attempted to hide him. He worked a series of odd jobs, attempting to disguise himself as a provincial Frenchman. For some months he worked as a horse groom, and then as an apprentice toolmaker for the French railroad. But he was never a very convincing tradesman. Missing the scholarly life, Mandelbrot clung to the few books he managed to find during this period, carrying them with him and reading whenever he had an opportunity—not the smartest move for someone trying to pass as a horse groom.

At one point, at least, Mandelbrot very narrowly escaped deportation—and likely execution. But mostly he managed to keep clear of German forces. His father had a closer call. As Mandelbrot would later tell it, his father was arrested during this period and sent to a nearby prison camp. Not long after, the prison was attacked by members of the resistance. The guards were neutralized and the prisoners were set free. But the resistance fighters were ill prepared to defend the camp, so they urged the prisoners to flee quickly to escape capture by German reinforcements.

Lacking a plan or a clear route to safety, the prisoners set out in a group on the road to Limoges, the nearest major town. Shortly after leaving the camp, however, the elder Mandelbrot realized that this was a disastrous idea: they were traveling in a large pack, moving in the open on a major road. Tracking them would be easy. The others couldn't be persuaded, so Mandelbrot's father left the group and struck out on his own. He headed toward a nearby forest, planning to slowly make his way back to where his family had been hiding before his arrest. As he moved through the wilderness, he heard a gut-wrenching noise: behind him, back at the main road, a German dive bomber had found the other prisoners.

Life during wartime is an unpredictable thing. In Thomas Pynchon's novel *Gravity's Rainbow,* one of the characters, Roger Mexico, is a statistician charged with keeping track of where the V-2 rockets land in London during the final days of the Third Reich. He finds that the rockets are falling according to a particular statistical distribution — the one you would expect if they were equally likely to fall anywhere in the city. Mexico is surrounded by people desperate to control their lives, to save themselves from the rockets' whimsical paths. To these onlookers, Mexico's charts and graphs hint at some underlying pattern, something they might use to predict where the next rocket will fall.

Some areas of the city seem to be hit quite often. Others, rarely. But to assume that these patterns say anything about where the *next* rocket will fall is to commit the same fallacy as the roulette player who is convinced that a particular number is "due." Mexico knows this. And yet he, too, finds the data seductive, as though the very randomness of the pattern holds the key to its power. And it does, at least if you happen to be standing on the street where the next rocket falls.

Yet mathematically, this sort of randomness is mild. The V-2 rockets were fired systematically, several a day, aimed roughly at London. Working out the odds of how many rockets would land on St. Paul's Cathedral or in Hammersmith was a lot like working out how many times a roulette ball would fall into red 25. Indeed, many of the situations we think of as random are like this. So many, in fact, that it's easy to fall prey to the idea that all random events are like coin tosses or simple casino games.

This assumption underlies much of modern financial theory. Think back to when Bachelier was imagining how stock prices would change over time if they underwent a random walk. Every few moments, the price would tick up or down by some small amount as though God were flipping a coin. Bachelier discovered that if this was a good approximation of what was happening, the distribution of prices would look like a bell curve, a normal distribution. Osborne of course pointed out that this wasn't quite right; really, you expect the prices to change by some fixed *percentage* each time God flips his coin, rather than some fixed amount. This modification led to the observations that rates of return should be normally distributed and prices should be log-normally distributed.

The normal distribution shows up in all sorts of places in nature. If you took the heights of all of the men in a given part of the world and plotted how many of them were 5 feet 6 inches, how many 5 feet 7 inches, and so on, you would get a normal distribution. If you used a thousand thermometers and tried to take your temperature with each of them, the results would look like a normal distribution. If you played a coin-flipping game in which you got a dollar every time the coin landed heads, and you lost a dollar every time it landed tails, the probabilities governing your profits after many plays would look like a normal distribution. This is convenient: normal distributions are easy to understand and to work with. For instance, if something is normally distributed and your sample is large enough, the sample's average value tends to converge to a particular number; white men, on average, are about 5 feet 9 inches, and unless you are ill, the thousand thermometers' readings will average 98.6 degrees Fahrenheit. Your average profits in the coin-tossing game will converge to zero.

This rule can be thought of as the law of large numbers for probability distributions — a generalization of the principle discovered by Bernoulli, linking probabilities to the long-run frequencies with which events occur. It says that if something is governed by certain probability distributions, as men's heights are governed by a normal distribution, then once you have a large enough sample, new instances aren't going to affect the average value very much. Once you have measured many men's heights in a given region of the world, measuring one more man won't change the average height by much.

Not all probability distributions satisfy the law of large numbers, however. The location of the drunken vacationer in Cancun does — he is taking a random walk, so on average, he will stay right where he started, just as the average profits from a coin-tossing game converge to zero. But what if instead of a drunk trying to walk to his hotel, you had a drunken firing squad? Each member stands, rifle in hand, facing a wall. (For argument's sake, assume the wall is infinitely long.) Just like the drunk walking, the drunks on the firing squad are equally liable to stumble one way as another. When each one steadies himself to shoot the rifle, he could be pointing in any direction at all. The bullet might hit the wall directly in front of him, or it might hit the wall 100 feet to his right (or it might go off in the entirely opposite direction, missing the wall completely).

Suppose the group engages in target practice, firing a few thousand shots. If you make a note of where each bullet hits the wall (counting only the ones that hit), you can use this information to come up with a distribution that corresponds to the probability that any given bullet will hit any given part of the wall. When you compare this distribution to the plain old normal distribution, you'll notice that it's quite different. The drunken firing squad's bullets hit the middle part of the wall most of the time — more often, in fact, than the normal distribution would have predicted. But the bullets also hit very distant parts of the wall surprisingly often — much, much more often than the normal distribution would have predicted.

This probability distribution is called a Cauchy distribution. Because the left and right sides of the distribution don't go to zero as quickly as in a normal distribution (because bullets hit distant parts of the wall quite often), a Cauchy distribution is said to have "fat tails." (You can see what the Cauchy distribution looks like in Figure 3.)

One of the most striking features of the Cauchy distribution is that it doesn't obey the law of large numbers: the average location of the firing squad's bullets never converges to any fixed number. If your firing squad has fired a thousand times, you can take all of the places their bullets hit and come up with an average value — just as you can average your winnings if you're playing the coin-flip game. But this average value is highly unstable. It's possible for one of the squad members to get so turned around that when he fires next, the bullet goes almost parallel with the wall. It could travel a hundred miles (these are *very* powerful guns) — far enough, in fact, that when you add this newest result to the others, the average is totally different from what it was before. Because of the distribution's fat tails, even the long-term average location of a drunken firing squad's bullets is unpredictable.

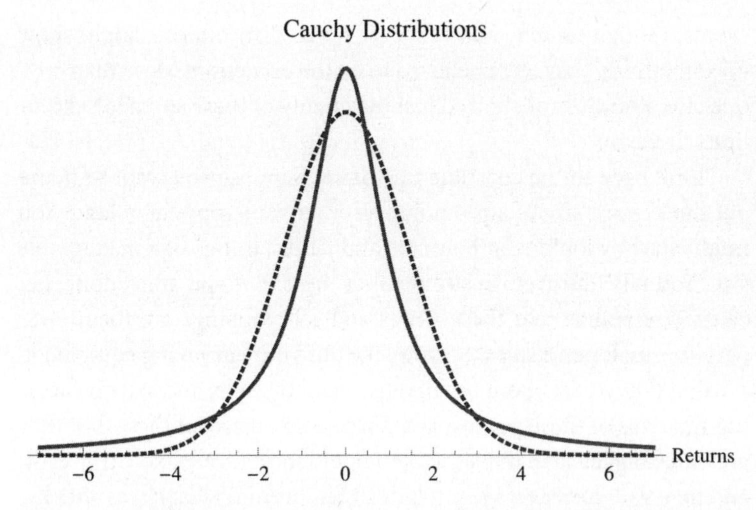

Figure 3: The location of a drunken vacationer trying to find his hotel room on a long corridor is governed by a normal distribution. But not all random processes are governed by normal distributions. Where the bullets fired by a drunken firing squad will land is determined by a different sort of distribution, called a Cauchy distribution. (Note that the *angle* at which the members of the drunken firing squad fire will be governed by a normal distribution; it's the location on the wall that they are trying to hit that is governed by the Cauchy distribution!) Cauchy distributions (the solid line in this figure) are thinner and taller than normal distributions (the dashed line) around their central values, but their tails drop off more slowly—which means that events far from the center of the distribution are more likely than a normal distribution would predict. For this reason, Cauchy distributions are called "fat-tailed" distributions. Mandelbrot called phenomena governed by fat-tailed distributions "wildly random" because they experience many more extreme events.

As Mandelbrot described it, the war, especially during the first two years under Vichy rule, left huge swaths of France unaffected for long periods. But then "a storm" would come through and wreak havoc, followed by another period of calm. So perhaps it is no surprise that Mandelbrot was fascinated by these bursts, by random processes that couldn't be tamed like a casino game. He called events that obeyed a Cauchy distribution *wildly* random, to distinguish them from the plain, mild randomness of the random walk, and he devoted much of his career to studying them. When Mandelbrot began his career, most statisticians assumed that the world is filled with normally distributed events; though Cauchy and other fat-tailed distributions might show up sometimes, they were believed to be the exception. More than anyone else, Mandelbrot showed just how many of these so-called exceptions there are.

Think back to the coastline of Britain. Suppose you want to figure out the average size of a promontory, or any outcropping of land. You might start by looking at boulders and jetties, things of a manageable size. You take the average size of all of these. But you aren't done, because you realize that these jetties and outcroppings are themselves parts of small peninsulas. So you take out your surveying equipment, sensing that you're about to fall down a rabbit hole, and start measuring the sizes of these peninsulas. There aren't many of these, but they are *much* bigger than the jetties and boulders you've looked at already, and now your new average is totally different from what it was after the first round of measurement. And what's more, you haven't even taken into account the still-larger structures, like Cornwall. Or the whole west coast of Britain itself, since from a geological perspective it's just an outcropping from mainland Eurasia. And while you're at it, you probably need to consider smaller structures, too. Why count boulders that are several feet across, but not rocks that are just a few inches across?

Each time you cast your net wider, the average changes dramatically. You can't seem to narrow in on a single figure. Dismaying as it is for our Sisyphean surveyor, there's no expected value for the average size of a feature on a coastline. This is a general property of fractals, following from their self-similarity. From one point of view, they are beautifully ordered and regular; from another, wildly random. And if fractals are everywhere, as Mandelbrot believed, the world is a place dominated by extremes, where our intuitive ideas about averages and normalcy can only lead us astray.

Though he never provided details, Mandelbrot often alluded to a particularly harrowing experience toward the end of 1943, while he was hiding with members of the French resistance. Afterward, his protectors realized that Mandelbrot couldn't remain in Tulle, and they secured a place for him as a postgraduate student at a preparatory school in Lyon.

Moving Mandelbrot was a risky proposition. Lyon was one of the most dangerous cities in southern France for both Jews and resistance sympathizers; Mandelbrot was both. Nikolaus Barbie, an SS officer, led the local Gestapo outpost from a hotel near the center of town. Known as the Butcher of Lyon, he was later convicted of war crimes for the deportation of nearly one thousand of the region's Jews. But Mandelbrot was not proving to be a very persuasive rural journeyman, and the resistance fighters who were caring for him needed a place where he wouldn't be so conspicuous. A school was a natural choice: Mandelbrot was the right age and he carried himself like a scholar. He would attend under an assumed identity and live in the dormitories. Yet even with a good cover, Mandelbrot couldn't risk venturing beyond school grounds. He was a prisoner as much as a student.

To complete the deception, Mandelbrot sat in on classes. But no one expected him to learn much. The school was designed to prepare the very brightest students for the difficult exams required for entrance to the *grandes écoles.* The atmosphere was often competitive and fast paced. Since Mandelbrot had not engaged in any academic work from the spring of 1942 to early 1944, when he enrolled at the school, he had once again fallen far behind his peers. It would be virtually impossible for him to catch up, given the caliber of his classmates and their ample head start.

At first, things went as expected. Mandelbrot sat quietly in the classes, pretending to be a student. He understood nothing. A week went by and then another. Mandelbrot listened as the instructor quizzed students on problems in abstract algebra, pushing them to compete to find the solutions as quickly as possible in preparation for the timed exams. Still, Mandelbrot understood nothing. He could guess at what the problems meant, but he had no clue how to solve them, and the discussions of various methods were lost on him. And then something remarkable happened. One day, after the teacher gave the class a problem to solve, an image appeared in Mandelbrot's mind. Without thinking, he raised his hand. Surprised, the teacher called on him. "Isn't this equivalent to asking whether these two surfaces intersect one another?" Mandelbrot asked, describing the two shapes he was picturing. The teacher agreed that the problems were equivalent but pointed out that the goal was to solve the problems quickly, not interpret them geometrically.

Mandelbrot sat back in his chair, silenced by the rebuke. But when the teacher read the next problem, he again tried to think of it in spatial terms. He very quickly saw what the shapes in question were. Soon he realized he could do this reliably. He had, it turned out, a "freakish" (his word) gift for visualizing abstract algebraic problems. But as his teacher reminded him, just coming up with a geometrical interpretation of a problem wouldn't help him on the test, and so Mandelbrot began thinking about how to put his talent to use. He didn't see a way to solve the problems using just his geometrical intuition, at least not in the way the teacher wanted. But he could very quickly guess what the answer had to be. And he was usually right. Soon, despite his poor preparation and unusual status, Mandelbrot was embraced by the school.

Liberation came in the summer of 1944. By the end of August, the Mandelbrots had moved back to Paris. Though he had been in Lyon for only six months, a single academic term, Mandelbrot's experience there changed the course of his life. He learned an enormous amount and discovered an unusual gift for geometry, but more importantly, he had reclaimed his education. He decided to continue his preparations for the *grande école* examinations, and in 1944, he was admitted to one of the most prestigious preparatory schools in Paris. After performing well on the exams, he gained entrance to several *grandes écoles,* including the most selective of all, the École Normale Supérieure.

He attended the École Normale Supérieure for two days before deciding that he couldn't bear life in an ivory tower. His time away from the academy had made him all too conscious of real-world problems. Mandelbrot immediately transferred to the more practical and scientifically oriented École Polytechnique. The choice augured Mandelbrot's path through academia: in each instance, faced with a choice between the pure and the applied, Mandelbrot chose the applied. In doing so, he brought his "freakish" geometrical gifts to bear on applied problems that had previously been overlooked, or that had seemed too difficult to crack. Like Bachelier before him, Mandelbrot asked questions that had never before occurred to anyone with his mathematical abilities — and he found answers that changed how scientists see the world.

Much later, Mandelbrot would attribute his remarkable career to two things. First was his unusual and oft-disrupted education. Mandelbrot ultimately made his way to a *grande école,* and on to a PhD. But the journey wasn't easy, and it forced him to be resourceful and independent in ways that he wouldn't have been had he followed a more traditional path. The second was a series of serendipitous discoveries that introduced him to various pieces of an intellectual puzzle. Zipf's formula, which he learned about when his uncle tossed a review in his face, was one such discovery. Another occurred several years later, soon after Mandelbrot finished graduate school.

At the time, he was working at IBM, another beneficiary of the industrialization of physics. Though he often expressed pride at having completed graduate school without an advisor, this didn't help when it came to seeking employment. He enjoyed a stint as a postdoctoral researcher at Princeton's Institute for Advanced Study and then spent some time back in Europe, working on thermodynamics for the French government's research center, CNRS. But a full-time faculty position proved elusive, and Mandelbrot's nascent disillusionment with the mathematical firmament deepened. When he at last received an offer from IBM in 1958 to work as a staff scientist for its research division, he jumped at the chance, even though, in his words, "there was no great distinction [in] getting an offer from IBM then."

One of the goals of IBM's research division was to find applications for its newest computers. Mandelbrot was assigned to work on economic data. His bosses hoped that if Mandelbrot could show how useful computers were for economics, banks and investment houses might be convinced to buy an IBM mainframe. In particular, he was looking at data describing income distributions throughout society. (Banks weren't necessarily interested in this specific question; rather, the idea was to use Mandelbrot's research as proof of concept, to demonstrate how efficient a computer could be at number-crunching financial data.)

Income distribution had been studied before, most famously by a nineteenth-century Italian engineer, industrialist, and economist named Vilfredo Pareto. A strong believer in laissez-faire economics, Pareto was obsessed with the workings of the free market and the accumulation of capital. He wanted to understand how people got rich, who controlled wealth, and how resources were doled out by market forces. To this end, he gathered an immense amount of data on wealth and income, drawing on such diverse sources as real estate transactions, personal income data from across Europe, and historical tax records. To analyze these data, Pareto would make elaborate graphs, with income levels and wealth on one axis, and the number of people who had access to that wealth on the other.

For all the diversity of his data sources, Pareto found a single pattern over and over again. As he described it, 80% of the wealth in any country, in any era, is controlled by 20% of the population. The pattern is now known as Pareto's principle, or sometimes the 80–20 rule. At the time, Pareto interpreted these results much as Zipf would have, as evidence for a "social law" revealing that wealth is not distributed randomly but rather by some mysterious force that shapes markets and societies. Once Pareto began looking, the law seemed to apply to everything. Eighty percent of a company's sales are usually due to just 20% of its customers. Eighty percent of crimes can be traced to just 20% of criminals. And so on. (Nowadays, Pareto's principle is seen to hold approximately in many places, such as in the ratio of health-care costs to patients in the United States.)

The most interesting thing about Pareto's work, at least from Mandelbrot's point of view, wasn't the idea that Pareto's data revealed some mathematical law of society. Instead, it was the particular relationship between the income distribution for a whole country and for a small portion of that country. Pareto showed that the 80–20 rule held, at least approximately, for a country as a whole. But what if you asked a slightly different question: How is income distributed among that 20% of the population that controls the overwhelming majority of wealth?

Remarkably, the same pattern emerges. If you look at just the wealthiest people in a country, 80% of *their* wealth is controlled by just 20% of them. The superrich tend to have the same disproportionate amount of wealth as the plain old rich. And indeed, the pattern continues. Eighty percent of the resources controlled by the superrich are consolidated in the hands of the *ultra*-superrich. And so on.

This kind of pattern should be familiar by now. Wealth distribution across a population displays a kind of self-similarity, or a fractal pattern. In fact, the distributions that Pareto discovered, called Paretian distributions, are a type of fat-tailed distribution — revealing a kind of wild randomness in income distribution, though not quite as wild as the drunken firing squad's shots. When Mandelbrot was looking at the data for IBM, he had not yet invented fractals. His seminal work on the coastline paradox was almost a decade away. But similarly to Pareto half a century before him, something about the pattern struck Mandelbrot. It reminded him of his doctoral work on Zipf, who also had discovered an odd self-similarity in how word frequencies were distributed.

Although Mandelbrot had largely left academia, his work for IBM on wealth distribution was of some interest to mainstream economists, and so he was occasionally invited to give scholarly talks. It was in 1961, immediately before one of these lectures, that he made his second serendipitous discovery.

The talk was to be delivered to Harvard's economics department. Shortly before it was scheduled to begin, Mandelbrot met with one of the faculty members, an economist named Hendrik Houthakker. As soon as he walked into Houthakker's office, Mandelbrot noticed a drawing on Houthakker's chalkboard. It was nearly identical to the graph that Mandelbrot was planning to use in his talk, as part of his discussion of income distribution and Pareto's principle. Mandelbrot guessed that Houthakker must have been working on a similar problem and made some comment about their shared interests. Houthakker responded with a blank stare.

After another awkward attempt or two, Mandelbrot realized that something was wrong. He backed up and pointed to the graph on the board. "Isn't that a wealth distribution plot?" Puzzled, Houthakker explained that the drawing on his board had been from a meeting with a graduate student earlier in the day, during which Houthakker and the student were discussing historical data on cotton prices. The picture was a graph of daily returns from cotton markets.

Houthakker went on to explain that he had been working on cotton markets for a while now, but the data weren't cooperating with theory. By this time, Bachelier's work had been rediscovered and economists had begun to accept that markets undergo a random walk, as Bachelier and Osborne had argued. Houthakker was interested in verifying this hypothesis by looking at historical data. If the random walk thesis was correct, you should see many small price changes over the course of a day or a week or a month, but very few large ones. What Houthakker's data showed, however, was not what the theory predicted: he was seeing too many very small changes, but also far too many very large ones. Worse, he was struggling to come up with a value for the average price change, as Bachelier's theory predicted must exist. Every time Houthakker looked at a new set of data, the average would change, often dramatically. In other words, cotton prices seemed to behave more like a drunken firing squad than a drunken vacationer.

Mandelbrot was intrigued. He asked Houthakker if he could look more closely at the data, and Houthakker agreed; in fact, Houthakker told Mandelbrot that he could have it all, since he was ready to abandon the project.

Back at IBM, Mandelbrot had a small team of programmers tear through boxes of Houthakker's cotton data, analyzing everything in detail. He quickly confirmed Houthakker's most troubling findings: it appeared that there was no "average" rate of return. The prices looked random, but they weren't explained by the standard statistical tools or Bachelier's and Osborne's theories. Something weird was going on.

Mandelbrot had seen unusual distributions before. In addition to studying Zipf's and Pareto's work, he was familiar with a third kind of distribution, discovered by one of his professors in Paris, Paul Lévy. It was Lévy who, upon reading a small section of one of Bachelier's papers, concluded that Bachelier's work was plagued with errors. Much later, Lévy would recognize his own mistake and apologize to Bachelier. Part of what made Lévy return to Bachelier's work was a renewed interest in random walk processes and probability distributions. Ironically, this later work of Lévy's received far less attention than his earlier work, leaving Lévy alienated and obscure at the twilight of his career.

Lévy's work on random processes had led him to study a class of probability distribution now called Lévy-stable distributions. The normal and Cauchy distributions are both examples of Lévy-stable distributions, but Lévy showed that there is a spectrum of randomness, ranging between the two. (In fact, there are even wilder varieties of randomness than the Cauchy distribution.) Wildness can be captured by a number, usually called alpha, that characterizes the tails of a Lévy-stable distribution (see Figure 4). Normal distributions have an alpha of 2; Cauchy distributions have an alpha of 1. The lower the number, the more wildly random the process (and the fatter the tails). Distributions that have alpha of 1 or less don't satisfy the law of large numbers — in fact, it isn't possible to even define the average value for a quantity that wild. Distributions with alpha between 1 and 2, meanwhile, have average values, but they don't have a well-defined average variability — what statisticians call volatility or variance — which means it can be very hard to calculate an average value from empirical data, even when the average exists.

Houthakker, trained as an economist, likely knew very little about Lévy's late work. But Mandelbrot had been a disciple of Lévy's. And so when he saw the detailed data from Houthakker, something clicked. Houthakker was right that cotton prices didn't follow a normal distribution — but they also didn't follow a Cauchy distribution. They were somewhere in between, with an alpha of 1.7. Cotton prices were random, all right — far more wildly random than Bachelier or Osborne could have imagined.

Cotton markets were the first place that Mandelbrot found evidence of Lévy-stable distributions. But if cotton prices varied wildly, he wondered, why should other markets be different? Mandelbrot quickly began collecting data on markets of all sorts: other commodities (like gold or oil), stocks, bonds. In every case he found the same thing: the

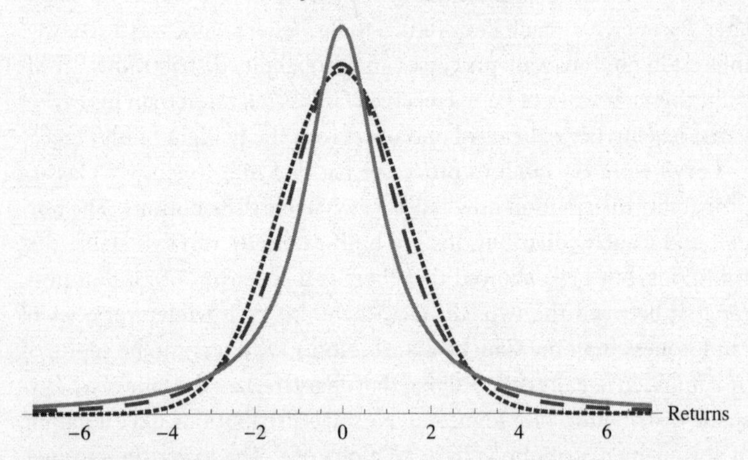

Figure 4: Normal distributions and Cauchy distributions are two extreme cases of a class of distributions called Lévy-stable distributions. Lévy-stable distributions are characterized by a parameter called alpha. If alpha = 2, the distribution is a normal distribution; if alpha = 1, it is a Cauchy distribution. Mandelbrot argued that real market returns are governed by Lévy-stable distributions with alpha between 1 and 2, which means that returns are more wildly random than Osborne had thought, though not as wild as a drunken firing squad. This figure shows three Lévy-stable distributions. As in Figure 3, the solid line corresponds to a Cauchy distribution and the dotted line is a normal distribution. But the third curve is a Lévy-stable distribution with alpha = 3/2. It's a little taller and a little narrower than a normal distribution, and its tails are a little fatter, but it's not so extreme as a Cauchy distribution.

alphas associated with these markets were less than 2, often substantially so. This meant that Bachelier's and Osborne's theories of random walks and normal distributions faced a big problem.

Mandelbrot made the connection between Pareto distributions and Lévy-stable distributions in 1960, the year after Osborne's first paper; he published the extension of this work to cotton prices in 1963, early enough that Paul Cootner, the MIT economist who edited the collection of essays that included Bachelier's and Osborne's work, was able to include a paper by Mandelbrot outlining his alternative theory. This meant that the volume that brought Bachelier's and Osborne's work to the wider community of economists and financial theorists already included hints that simple random walk models were not the whole story. Around 1965, financial theorists had a choice, though it surely didn't feel that way to them at the time: they could follow Osborne and others who showed how traditional statistical methods, developed largely in the context of physics, could be used to analyze and model stock market returns; or they could follow Mandelbrot, who showed that despite this remarkable power, there was reason to think the traditional methods had shortcomings. Weighing in on the traditionalists' side was the fact that the older methods were better understood and simpler. Mandelbrot, meanwhile, had some highly suggestive data on his side.

The field chose Osborne. Cootner made the argument this way at a 1962 meeting of the Econometric Society,* in response to Mandelbrot's work on cotton prices:

> Mandelbrot, like Prime Minister Churchill before him, promises us not utopia but blood, sweat, toil, and tears. If he is right, almost all of our statistical tools are obsolete. . . . Almost without exception, past econometric work is meaningless. Surely, before consigning centuries of work to the ash pile, we should like to have some assurance that all our work is truly useless.

Much of the field took a similar view. At this point, the (mild) random walk hypothesis was still young, but a growing number of researchers, Cootner included, had already staked their careers on it. It is easy to see Cootner's remarks as a reactionary attempt to fend off a young researcher who had caught out the errors of the (recent) past. Surely Mandelbrot saw it this way, and perhaps we all should now that many practitioners and theorists alike have recognized the importance of fat-tailed distributions. For instance, some people — most notably, Nassim Taleb, a hedge fund manager and professor at Polytechnic Institute of New York University who wrote an influential book called *The Black Swan,* as well as Mandelbrot himself — have recently argued that finance took a wrong turn in 1965 by continuing to assume mild randomness when really financial markets are wild.

---

* Econometrics is the statistical study of economic data, including but not limited to finance.

But that argument misses an important point about the way the science of finance has developed. In the 1960s, traditional statistics was a mature field with an enormous toolbox. Mandelbrot was coming forward with little more than a suggestion and a few pictures. It would have been essentially impossible to do the kind of work that Osborne, Samuelson, and many others working in finance and econometrics did during this period without the tools of traditional statistics. Mandelbrot's project simply wasn't well enough understood. It would be like telling a carpenter that screws are much stronger than nails, when the carpenter has a hammer and no one has yet invented the screwdriver. Even if the house would be stronger if built with screws, you'd still get much farther working with a hammer and nails, at least for a while.

For this reason, pushing forward with the simpler available tools while Mandelbrot and his early converts worked out the consequences of his work on fractals and self-similarity was the only sensible choice. What the field implicitly understood is that you need to start with the simplest theory that works, get as far as you can, and then ask where the theory you've built has gone wrong. In this case, once you have established that stock market prices are random (at least in some sense), the next step is to assume that they are random in the simplest possible way: that they just follow a random walk. This is what Bachelier did. Osborne then pointed out that this couldn't be right, since it would mean that stock prices could become negative, and so he complicated the model ever so slightly by suggesting that market rates of return follow a random walk. He then showed that this suggestion explained the data much better than Bachelier's model.

Then came Mandelbrot, who said that Osborne's suggestion wasn't quite right either, because if you looked at price data in detail, you would see a different pattern from the one Osborne thought he had found. Not dramatically different, though; the pattern Mandelbrot identified doesn't say that prices aren't random, but that prices are random in a slightly different way from what Osborne had believed. The differences between Osborne's model and Mandelbrot's can hardly be dismissed, but they become important only in the context of extreme events. On a typical day, there aren't going to be any extreme events (according to either theory), and so you usually won't notice much of a difference between the two models.

For this reason — as we will see in the next several chapters — when it came time for economists interested in financial markets to try to extend the ideas presented in Cootner's book, to put the randomness of stock market prices to work by using statistics to predict derivatives prices or to calculate the amount of risk in a portfolio, they had to pick between the simple theory that gave good results the vast majority of the time and the more cumbersome one that better accounted for certain extreme events. It made perfect sense to start with the simpler one and see what happened. If you make good assumptions, if you idealize effectively, you can often solve a problem that otherwise couldn't be solved — and get a solution that is quite close to correct, even if some of the details are wrong. Of course, all along, you know you've made assumptions that aren't quite right (markets are not perfectly efficient; returns and not prices follow a simple random walk). But they're a start.

It is also too simple to say that Mandelbrot was ignored in the decades immediately following his early papers on cotton. Most economists followed Osborne's lead when building on the randomness of markets to study related topics. But a dedicated core of mathematicians, statisticians, and economists put Mandelbrot's proposals to the test with ever more detailed data, and ever more sophisticated mathematical methods — most of which were developed specifically for the purpose of better understanding what it would mean if the world were really as wildly random as Mandelbrot said. This work confirmed Mandelbrot's basic thesis, that normal and log-normal distributions are insufficient to capture the statistical properties of markets. Rates of return have fat tails.

That said, there's a wrinkle in the story. Mandelbrot made a very specific claim in his 1963 papers: he said that markets were Lévy-stable distributed. And except for the normal distribution, the volatility of Lévy-stable distributions is infinite, which means that most standard statistical tools don't apply for analyzing such distributions. (This is what Cootner was alluding to when he said that if Mandelbrot was correct, the standard statistical tools were obsolete.) Today, the best evidence indicates that this *specific* claim, regarding infinite variability and the inapplicability of standard statistical tools, is false. After almost fifty years of research, the consensus is that rates of return are fat-tailed, but they aren't Lévy-stable. If this is correct, as most economists and physicists working on these topics believe it is, then the standard statistical tools *do* apply, even though the simplest assumptions of normal and log-normal distributions do not. But evaluating Mandelbrot's claims is an extremely tricky business — mostly because the important differences between his proposal and its nearest alternatives apply only in extreme cases, data for which are very hard to come by. And even today, there is disagreement about how to interpret the data we do have.

The fact that Mandelbrot's claims were likely too aggressive makes his legacy a little more difficult to evaluate. Some writers today insist that Mandelbrot was never given his due, and that a proper appreciation of his ideas would solve all the world's problems. While this is not entirely true, a few things are certain. Extreme events occur far more often than Bachelier and Osborne believed they would, and markets are wilder places than normal distributions can describe. To fully understand markets, and to model them as safely as possible, these facts must be accounted for. And Mandelbrot is singularly responsible for discovering the shortcomings of the Bachelier-Osborne approach, and for developing the mathematics necessary to study them. Getting the details right may be an ongoing project — indeed, we should never expect to finish the iterative process of improving our mathematical models — but there is no doubt that Mandelbrot took a crucially important step forward.

After a decade of interest in the statistics of markets, Mandelbrot gave up on his crusade to replace normal distributions with other Lévy-stable distributions. By this time, his ideas on randomness and disorder had begun to find applications in a wide variety of other fields, from cosmology to meteorology. These fields were closer to his starting point in applied mathematics and mathematical physics. He remained affiliated with IBM for his entire career; in 1974, he was named an IBM

Fellow, which gave him considerable freedom to identify and develop his own projects, much like an academic researcher.

Gradually, as his ideas percolated through the many different scientific disciplines, Mandelbrot began to receive some recognition for his work. The book that introduced the term *fractals* to the wider world went through several revisions beginning in 1975 and culminated in *The Fractal Geometry of Nature* in 1982. It was a cult sensation, and it turned Mandelbrot into a semi-public figure. By the early 1990s, he had collected a long list of significant honors and awards, including election to the French Legion of Honor in 1990 and the Wolf Prize for physics in 1993. In 1987, he began teaching mathematics part-time at Yale — ultimately receiving his first tenured faculty position in 1999, at the age of seventy-five. He continued to lecture and work on original research, right up to his death, on October 14, 2010.

In the early 1990s, Mandelbrot sensed that the moment had arrived to move back into finance, and this time he had more success. Over the previous three decades, his ideas had developed and matured — benefiting greatly from their application to other fields — and so when he returned to thinking about economics, he had a much larger set of mathematical tools on which to draw. Meanwhile, markets had changed, so that far more practitioners on Wall Street and elsewhere were equipped to understand and incorporate Mandelbrot's ideas. It was at this point that the recognition of fat-tailed distributions reached the financial mainstream. But I am getting ahead of the story. It would take a blackjack sharp and a dilettantish ex-physicist to move finance to a place where it could take advantage of the insights of Bachelier, Osborne, and, ultimately, Mandelbrot.

CHAPTER 4

Beating the Dealer

THE YEAR IS 1961. The place, Las Vegas. It's a Saturday night in the middle of June. The temperature is hovering around 100 degrees even though the sun has already set. Inside the casinos, no one cares. Vegas is at the height of its postwar golden age. A dozen world-class resorts, the first of their kind, line the nascent Strip, from the Sahara in the north to the Tropicana in the south. The loud, smoky casino floors are packed with tourists from across the country hoping to get lucky at the tables, or at least ogle some celebrities. This is the Vegas of the original *Ocean's Eleven*, the Vegas of Michael Corleone, the Vegas James Bond visits in *Diamonds Are Forever*. The Vegas of Elvis and the Rat Pack, Liberace and the Marx Brothers.

A slender man with a crewcut, just shy of thirty, is sitting at a roulette table. He stares straight ahead, his face impassive behind a pair of horn-rimmed glasses. The crowd packs in around him, boisterously throwing chips at the table. But he ignores them. He looks intent, deeply focused, though on what is unclear. The minutes tick by and the crowd starts to wonder if he's forgotten about the game. Then, at the last possible moment, he places his chips on seemingly random spots on the board. For one round, it's black 29, red 25, black 10, red

27. On the next, it's black 15, red 34, black 22, and red 5. To the people around him he seems crazy. Roulette players often have systems, but they're consistent, like lottery players: you bet your birthday, or your girlfriend's phone number. Or, if you like a safer bet, you play a color. But this guy's bets keep changing, as though someone is whispering the future into his ear. Whatever he's doing, it doesn't seem quite right. Especially because he's winning. A lot.

His name is Edward Thorp. Today he is one of the most successful hedge fund managers in history. In June 1961, he was only a few years out of graduate school. He had just been hired as an assistant professor of mathematics at New Mexico State University. In graduate school he had specialized in the mathematics of quantum physics. But Thorp was also fascinated by games. He was particularly interested in strategy games: blackjack, poker, baccarat. Even the ancient Chinese game Go. But on that sweltering Vegas night back in 1961, he was playing roulette. This was odd, because the results of spinning a roulette wheel should be perfectly random. Each spin is independent of the spin before and the spin after. There's no place for strategy.

Back at the roulette table, a man and a woman walk past Thorp, gulping whiskey sours. A cheer goes up at another table as someone from Des Moines wins big. Distracted for a moment, Thorp looks up —just in time to catch a look of horror from the woman next to him. Thorp's hand shoots to his ear. Attracted by the movement, a few bystanders glance in his direction and catch a glimpse of . . . what is that? An *earpiece?* Thorp is already on his feet, gathering his chips and stuffing them into his pockets with one hand while his other hand remains pinned to his ear. He pushes his way out of the crowd and hurries toward the street.

We've seen how Bachelier and Osborne used insights from physics to propose that markets can be understood in terms of a random walk, and how Mandelbrot refined that idea. Their work revolutionized the study of financial markets, once economists came to appreciate it. But all three were firmly ensconced in academia. Bachelier worked at the Bourse, but there's no evidence that he put his ideas to any use there, and he certainly never made much money. Osborne may have turned to finance in an attempt to feed his family, but he ultimately concluded that there was no profit to be had in speculating on the unrelieved bedlam of financial markets. Mandelbrot, too, seems to have avoided trading.

Certainly ideas introduced by Bachelier, Osborne, and Mandelbrot percolated through economics departments and affected how traders thought about financial markets. For instance, the 1973 book *A Random Walk Down Wall Street,* by Princeton economist Burton Malkiel, has become a classic among investors of every stripe; it owes a great deal to Osborne in particular, though this influence is largely uncredited.

But the introduction, and subsequent sharpening, of the random walk hypothesis is only part of the story of how physicists have shaped modern finance. Physicists have been equally, or even more, influential in their role as practitioners. Ed Thorp is a prime example. He accomplished what Bachelier and Osborne never could: he showed that physics and mathematics could be used to profit from financial markets. Building on the work of Bachelier and Osborne, and on his own experience with gambling systems, Thorp invented the modern hedge fund — by applying ideas from a new field that combined mathematical physics and electrical engineering. Information theory, as it's known, was as much a part of the 1960s as the Vegas Strip. And in Thorp's hands, it proved to be the missing link between the statistics of market prices and a winning strategy on Wall Street.

Thorp was born at the peak of the Depression, on August 14, 1932. His father was a retired army officer, a veteran of the First World War. When Thorp was born, his father was fortunate enough to have found work as a bank guard, but money was still tight and the young Thorp developed an early instinct for thrift and financial savvy. He realized he could buy a packet of Kool-Aid mix for a nickel but could make six glasses with each packet. So he sold glasses of cold Kool-Aid to WPA workers for a penny each. He bet a storekeeper that he could add up a tab in his head faster than the cash register and won himself an ice cream cone. An older cousin showed him that the slot machines at his local gas station were rigged so that if you jiggled the handle right, they would pay out.

When World War II began, the Thorps headed west to find work in defense manufacturing. They settled down in Lomita, California, just south of Los Angeles. Both parents took jobs, leaving Thorp to fend for himself. It was around this time that he discovered something even more exciting than betting on his quick head: blowing stuff up. He started with a children's chemistry set, a gift from his parents, and ultimately set up a junior mad scientist's lab in the garage. While his parents helped with the war effort, Thorp was building pipe bombs and blowing holes in the sidewalk with homemade nitrocellulose. Later his tinkering would expand to include playing with telescopes and electronics, including ham radios.

Thorp's boyhood penchant for explosives belied a deep fascination with the science behind his experiments, and along the way he learned a considerable amount of chemistry and physics. In 1948, at the end of his sophomore year in high school, Thorp signed up to take an All Southern California test in chemistry, competing for a scholarship to the University of California. When he told his chemistry teacher of his plan, the teacher was dubious. Thorp was over a year younger than the other competitors, who were preparing for college. But after the teacher gave Thorp a practice exam, he was convinced. Thorp didn't know everything, but he had clear aptitude. Thorp's teacher recommended three books for Thorp to read and gave him a stack of practice tests to work on over the summer.

When the test results came back, Thorp learned that he had come in fourth overall. The results were remarkable, but he knew he could do better. The version of the test he took included a new section that hadn't been on the previous year's test, and it had called for a slide rule. Thorp had a ten-cent slide rule, small and poorly machined. The numbers didn't always line up correctly, introducing errors in Thorp's calculations. Thorp was convinced that if he'd had a proper slide rule, he would have won the competition. The problem was that he couldn't take the chemistry test again. So the following year he signed up for the corresponding test in physics. This time he came in first and won the scholarship, which paid his way through UCLA. He'd successfully parlayed backyard explosives into college tuition.

Since it was physics rather than chemistry that had gotten Thorp to UCLA, he decided to make it his major. Four years later, he stayed on for graduate school. Thorp loved his studies, but graduate school wasn't a natural choice for him, given his lack of means. If not for the scholarship competition, it's unlikely that he would have been able to afford college. And now, when he was twenty-one, money was as big an issue as ever. Thorp mustered a budget of $100 a month — about $850 in 2012 dollars — half of which went immediately to rent. Strapped for cash, Thorp began scheming about ways to make a little extra money on the side, à la his childhood exploits.

It was a conversation on just this topic — how to make extra money without much work — that first got Thorp thinking about roulette. It began as a debate at the UCLA Cooperative Housing Association dining room in the spring of 1955, just as Thorp was preparing to finish his master's degree in physics. The first Las Vegas casinos had just begun to open, and gambling was a hot topic. One of Thorp's friends suggested that gambling was a good way to get rich quick. The problem, someone else pointed out, was that you usually *lose*. After a discussion of whether it was possible to get an advantage at various games (that is, improve your chances so you win more often than you lose), roulette came up. Most of Thorp's colleagues argued that roulette was a terrible choice for a get-rich-quick scheme. Maybe if the wheel had something wrong with it, certain numbers would come up more often than others. But the wheels at big casinos, like the ones in Las Vegas or Reno, were made so precisely that you could never find an imperfection to exploit. Roulette wheels were as close to random as you could get, and without some special trick, the odds were against you.

Thorp didn't disagree with the premise. But he thought the conclusion was wrong. After all, he reasoned, physicists are good at predicting how things like wheels behave. If a roulette wheel really is perfect, well, then shouldn't normal high school physics be enough to predict where a ball starting at such and such a place, rolling around a wheel spinning with such and such velocity, would land? You don't need quantum physics or rocket science to figure out how balls roll around wheels. The fact that roulette wheels are so perfectly manufactured could only help: there aren't going to be small imperfections in the wheel that might throw off your calculations, and each wheel should be pretty similar to every other.

To test his hypothesis, Thorp started doing experiments. He did a few calculations and then bought a cheap, half-size wheel and filmed a ball going around it so he could watch, frame by frame, how it behaved. Meanwhile, he thought about how to put his idea to use. Major casinos accept bets even after the ball is moving, so in principle it's possible to know the initial speed and position of the wheel and ball, which ought to be all you need to calculate where the ball will land, *before* you make your bet. He fantasized about building a machine that could quickly make the necessary calculations. But he didn't get very far. Vegas roulette wheels might be flawless, but the toy wheel he bought was a piece of junk. Watching the films convinced him that the wheel was useless for his experiments; professional wheels, meanwhile, cost well over $1,000 — an impossible investment for an impoverished grad student.

So Thorp gave up on roulette, at least for a while. After finishing his master's degree, he began working on his doctorate, again in physics. He quickly realized, however, that his mathematical background wasn't sufficient to tackle the newest topics. He made a list of the courses he would need to take, most of which were in a then-burgeoning field known as functional analysis, and discovered that if he took them all, he'd have enough for a PhD in mathematics, while his work on physics would have just begun. And so he switched to math. All the while, his ideas about the physics of roulette spun around in his mind. He was sure that with the right resources — a professional roulette wheel and some computer know-how — he could strike it rich.

Soon after finishing his PhD, Thorp was awarded the prestigious C.L.E. Moore instructorship in mathematics at MIT — a position held a decade earlier by John Nash, the pioneering mathematician profiled by Sylvia Nasar in her book *A Beautiful Mind*. Thorp and his wife, Vivian, left Southern California and moved to Cambridge, Massachusetts.

They spent only two years on the East Coast before moving back west, to New Mexico. But it was enough to set their lives on a different track: it was at MIT that Thorp met Claude Shannon.

Shannon may be the only person in the twentieth century who can claim to have founded an entirely new science. The field he invented, information theory, is essentially the mathematics behind the digital revolution. It undergirds computer science, modern telecommunications, cryptography, and code-breaking. The basic object of study is data: bits (a term Shannon coined) of information. The study of things such as how light waves move through air or how human languages work is very old; Shannon's groundbreaking idea was that you could study the information itself—the stuff that's carried by the light waves from objects in the world to your retinas, or the stuff that passes from one person to another when they speak—independently of the waves and the words. It is hard to overstate how important this idea would become.

Information theory grew out of a project Shannon worked on during World War II, as a staff scientist at Bell Labs, AT&T's research division in Murray Hill, New Jersey. The goal of the project was to build an encrypted telephone system so that generals at the front could safely communicate with central command. Unfortunately, this was hard to do. There is only one code system that can be mathematically proven to be unbreakable. It's called a one-time pad. Suppose you start with a letter that you want to send to your friend but that you don't want anyone else to read. Say the letter has 100 characters in it, including spaces. To protect the letter with an unbreakable code, you need to come up with a random list of 100 numbers (corresponding to the number of characters in the letter) called a key, and then "add" these numbers to the characters in the letter. So if the first character in the letter is *D* (for "Dear John," say), and the first number in your random list is 5, you want to add 5 to *D* by moving down the alphabet by five letters. So you write down *I* as the first letter of the coded message. And so on. In order to decrypt the letter, your friend needs to have a copy of the key, too, which can then be used to subtract the right number from each letter and recover the original message. If the key is really random, there's no way to decrypt the encoded message without access to the key, since the randomness of the key will wash out any patterns in the original message.

A one-time pad such as I just described can be tricky in practice because the sender and the receiver have to have the same random keys. But in principle, the idea is simple. It gets more complicated when you try to implement the idea of a one-time pad for a *telephone* conversation. Now there are no letters to add a number to or subtract a number from. There are sounds, and what's more, the sounds are transmitted over long distances by a wire (or at least they were in 1944), which means that anyone who can gain access to the wire, at any point between the generals in the field and their home base, can listen in on the conversation.

The Bell Labs team realized that the essence of the one-time pad was the fact that patterns in the "signal," the message being transmitted, get lost amid the randomness of the "noise" — the key consisting of random numbers. So you need to take whatever medium is being used to carry the message (in this case sound) and add something to it that's totally random so that you can't make out any of the message-bearing patterns. In a telephone conversation, the word *noise* isn't a metaphor. Imagine trying to talk to someone with a loud vacuum cleaner running in the background. You wouldn't be able to make out much, if anything, of what the person was trying to say. This is the principle behind SIGSALY, the system that Shannon and his collaborators invented. If you add enough noise to whatever your general is saying, you can make it incomprehensible. Meanwhile, if you have access to a recording of the exact same random noise on the other side of the message, back in Washington, you can "subtract" it from the coded message to recover the original voice. Implementing the system was an engineering marvel: signal processing of the sort necessary to remove noise from a telephone line, even if you knew exactly what the noise sounded like, was only at its earliest stages. But Shannon and his team figured out how to make it work. SIGSALY devices were built at the Pentagon for Roosevelt, in Guam for MacArthur, in North Africa for Montgomery, and in the basement of Selfridges department store in London for Churchill.

Thinking about the relationship between a signal and noise led

Shannon to his most important insight — the basic idea underlying all of information theory and, by extension, the information revolution. Suppose you're driving on the highway, having a conversation with the person in the passenger seat. You're chatting away, and then an eighteen-wheeler passes by, and for a moment your passenger can hear only every other word you say because the truck is so loud. Will the passenger figure out what you were trying to say? It depends. Maybe you've just gotten started on your regular rant about traffic in Los Angeles. You complain about it constantly, so your friend knows the riff by heart. Just a few words — maybe "construction" or "bad drivers," plus an obscenity or two — would be enough to transmit the full force of your views on traffic. In fact, the passenger could be a complete stranger; no one likes traffic, and so a keyword here or there would be sufficient to get your message across. But what if you were trying to explain the details of a new film you just saw? Then every word could be important. Your passenger would have little idea what to make of it if all he could hear was "The lead — was — in the green —."

Shannon concluded that the amount of information carried by a signal has something to do with how easy it is for the receiver to decode, or in other words, on how unpredictable the signal is. Your rant on traffic doesn't contain much information — it's easy to predict; your film synopsis contains more. This is the essence of Shannon's information theory.

Perhaps the easiest way to see why this way of looking at information makes sense is to turn Shannon's picture around. Information is the kind of thing that takes you from feeling not so sure about something to feeling more sure about it. If you gain information, you *learn* something about the world. Now imagine two cases. Suppose you begin by thinking that the Yankees have a great chance of winning half their games in any given year, but that there's very little chance that there are aliens living on the moon. Shannon's essential insight could be put as follows: if you were to learn, as in become absolutely certain, that there *are* aliens living on the moon, you would have gained a lot more information than if you were to learn that the Yankees *have* won more than half their games this year. The reason? In Shannon's terms, it's that the probability of there being aliens on the moon is much, much lower than that of the Yankees (or any other team) winning half their games. This connection between the probability of a message and the information contained in the message provides the crucial link needed to quantify information. In other words, by connecting information with probability, Shannon discovered a way to assign a number to a message that measures the amount of information it contains, which in turn was the first major step in building a mathematical theory of information.

The invention of information theory turned Shannon into an overnight sensation, at least in the worlds of electrical engineering, mathematics, and physics. The applications proved to be endless. He stayed at Bell Labs for another decade after the war, before he moved to MIT in 1956.

Thorp arrived in Massachusetts in 1959, just a year out of graduate school. By then, Shannon held an endowed chair, with dual appointments in the mathematics and electrical engineering departments. His most important work had already been published and its influence was spreading rapidly. By the late 1950s, he was an academic rock star. Already famously eccentric, Shannon was now powerful enough to dictate his own terms to MIT: whom he would meet with, what he would teach, how much time would be devoted to research. He was not the kind of man whose office you would casually stick your head into — especially if you were just a lowly instructor. To meet Shannon, Thorp needed an appointment. And to get an appointment, he needed something worth talking about; as Shannon's secretary would later inform Thorp, Professor Shannon didn't "spend time on topics (or people) that didn't interest him."

Fortunately, Thorp had a topic that would entice Shannon. A few months before moving to Massachusetts, the Thorps had visited Las Vegas for the first time. They chose Vegas because they expected it to be a bargain: close to Los Angeles, plenty of inexpensive hotels, a lot to see and do. Plus, Thorp thought, he'd have a chance to scope out professional-level roulette wheels. But as it turned out, roulette wasn't Thorp's principal interest on this trip. Shortly before the young couple left for their vacation, a colleague passed along a recent academic article from the *Journal of the American Statistical Association.* It concerned the game of blackjack, or twenty-one.

As far as casino games go, blackjack is old — older, even, than roulette. Cervantes, the author of *Don Quixote,* used to play a variation in Spain in the early seventeenth century and wrote stories in which his characters became proficient at cheating. The game is typically played with one or more standard decks of cards. You start by placing your bet. The game begins with each player (including the dealer) being dealt two cards, and then players have a chance to ask for additional cards until they decide they've had enough or they "bust," which happens if their cards sum to more than twenty-one points. Number cards are worth their face value; face cards are worth ten points; and an ace can be worth either one point or eleven points, at the player's discretion. The goal is to have the highest number of points without going over twenty-one. At a casino, each player is competing individually against the dealer, who represents the house. The goal, then, is to beat the dealer without busting. If you win, the game pays a dollar for every dollar you bet unless your initial two cards add up to twenty-one. In that case, the game pays a $1.50-per-dollar bet.

Casinos always employ the same strategy. The dealer has to take a new card as long as his total number of points is less than seventeen. If it's seventeen or more, the dealer stops. And if the dealer busts, everyone wins. The twist, at least in a casino, is that although the players' cards are all dealt face up, one of the dealer's cards is dealt face-down, so the players do not get to see it until the end of the game. Not knowing what you're up against makes it more difficult to know when to stop asking for new cards.

Casinos have run blackjack tables for a long time. And they've made money doing it. This suggests, but doesn't *quite* prove, that the odds are with the house. The reason it doesn't quite prove it is that blackjack, unlike roulette, is a strategy game. The player has a choice to make: When do you ask for additional cards? Even by the early 1950s, as gambling took hold in Vegas, no one knew if there was a strategy that a player could adopt that would give him an advantage over the house. All anyone knew for sure was that whatever most people were doing, it was good for the house. Figuring out more than that would prove incredibly difficult. It involved calculating the probabilities of all of the possible hands, under all sorts of different circumstances. Millions of calculations.

This was just what a group of army researchers set out to do, beginning in 1953. Over the course of three years, using "computers" (which in the early 1950s meant people, perhaps with electronic adding machines), the army team worked out (almost) all of the possible hands, figured out their probabilities, and then devised what they claimed was the "optimal" blackjack strategy. It was this strategy that they published in the *Journal of the American Statistical Association,* and that Thorp decided to try on his trip to Vegas. It wasn't a winning strategy. According to the army's calculations, the house had an advantage even if you played with their optimal strategy, because of the essential role of uncertainty about the dealer's hand in the player's decision making. But the advantage was tiny. If you made a thousand one-dollar bets at successive hands of blackjack using their strategy, the army predicted, you should expect to have (on average) about $994 left at the end of the day. Compare this to slots, where you could expect to have about $800 left, and the optimal blackjack strategy looked pretty good. Unfortunately, the strategy wasn't simple, so Thorp had to make a cheat sheet; he wrote out all of the possibilities on a little card, which he consulted as he played.

He lost. Quickly. Starting with a pile of $10, Thorp was down to $1.50 within the hour. But the other people at the table lost even more quickly, and by the time Thorp left the table, he was convinced that the army's researchers were on to something. He was also convinced that he could do better.

The problem with the army strategy, as Thorp saw it, was that it treated each round of blackjack as independent: it was as though each time around, a brand-new deck was being used. But in real life, especially in 1958 (casinos have since changed the rules slightly), this wasn't the case. A dealer would shuffle a deck and then keep playing as long as there were enough cards to go around. This changes everything. Consider that the probability of receiving, say, an ace from a new deck is 4/52, since there are 4 aces in a deck of 52 cards. But suppose you're on your second hand, and on the first hand 10 cards came up, two of which were aces. Now the odds of getting an ace are 2/42, which is much less than 4/52. The point is that if your strategy depends on the probabilities of getting different card combinations, and if you're being careful, you need to take into account what cards have already been played. Adopting such a strategy, where you keep track of what cards have already been played and vary your strategy accordingly, is called card counting.

Card counting, Thorp believed, could make the odds in blackjack even better than what the army researchers found. Using MIT's IBM 704, one of the first mass-produced electronic computers, Thorp managed to prove that the player would have an advantage if he combined a modified version of the army's strategy with a simple card-counting technique. This was Thorp's in with Shannon. He wrote a paper describing what he had found, with the hope that Shannon would help him publish it.

When the day of the meeting arrived, Thorp knew the pressure was on. He had his thirty-second elevator pitch ready: what he wanted; why Shannon should care.

As it turned out, Thorp had little to worry about. Shannon immediately saw what was interesting about Thorp's results. And after a few piercing questions, Shannon was convinced that Thorp was the real deal. He made some editorial suggestions and suggested that Thorp tone down the title (from "A Winning Strategy for Blackjack" to "A Favorable Strategy for Twenty-One") and then offered to submit Thorp's paper to the *Proceedings of the National Academy of Sciences,* the most prestigious academic journal that would consider publishing such work (only members of the Academy could submit papers). Then, as Thorp prepared to leave, Shannon casually asked if Thorp had any other gambling-related projects. This kind of math, with clear and fun applications, was right up Shannon's alley. After a pause, Thorp leaned forward. "There is one other thing," he began. "It's about roulette . . ."

It was dusk on a snowy winter's evening in Cambridge, Massachusetts. A dark sedan circled the block once and then slowed to a stop in front of the Thorps' apartment building. The doors opened, and from each side of the car a beautiful young woman emerged. Both women had mink coats draped over their shoulders. They stepped back from the car to reveal its third passenger, a short man in his early sixties. His name was Manny Kimmel. He was the owner of a growing parking lot and funeral home concern known as the Kinney Parking Company. The Kinney Parking Company was in the process of going public. Over the next decade, under the joint leadership of Kimmel's son Caesar and legendary CEO Steve Ross, Kinney would rapidly expand: first to commercial cleaning and facilities management, and then to media. In 1969, Kinney Parking Company would acquire Warner Brothers Studios as the first step in a transformation that would ultimately culminate in Time Warner, which is today the world's largest media conglomerate.

In 1961 all of this was in the future. But Kimmel was already a very wealthy man. His fortune had been made the old-fashioned way: gambling and booze. Legend has it that Kimmel won his first parking lot, on Kinney Street in Newark, New Jersey, in a high-stakes craps game. And the early success of the Kinney Parking Company had as much to do with Kimmel's side business of running limousines to illegal gambling houses as it did with people parking their cars. During Prohibition, he teamed up with his childhood friend, the Jewish mobster Longy Zwillman. Zwillman would import rye whiskey from Canada and then use Kimmel's New Jersey garages to store it.

It was gambling that brought Kimmel to Thorp's doorstep that cold Sunday in February. A few weeks before, Thorp had given a public talk on his National Academy paper at the American Mathematical Society's annual meeting, in Washington, DC. This time around, he permitted himself a provocative title: he called the talk "Fortune's Formula: A Winning Strategy for Blackjack." Blackjack aside, Thorp's talk was a winning strategy for attracting media attention. He delivered the talk to a packed audience, and soon the AP and other news outlets came knocking. Within days, stories had begun to appear in the national media, including the *Washington Post* and *Boston Globe*. The dry annual AMS meeting rarely attracted much notice in the news, but something about an MIT mathematician taking Vegas to the cleaners struck a chord.

At first, Thorp reveled in the attention. His phone began ringing off the hook, with reporters looking for interviews and gambling fanatics hoping to learn Thorp's tricks. He boasted to reporters that if he could get sufficient funding for a trip to Vegas, he would prove that his system worked in practice. As a publicity stunt, the Sahara, one of the big Vegas Strip casinos, offered him free room and board for as long as he liked — trusting that Thorp's system, like the hundreds that preceded it, was at best a fantasy. But the Sahara wouldn't front Thorp gambling money, and on his $7,000-a-year salary, Thorp couldn't raise sufficient funds himself. (Since casinos have minimum bets, an early losing streak can wipe you out if you don't have a pile of cash on hand — even if you're very likely to win in the long run.)

This is where Kimmel came in. Some men like fine wines or expensive cigars. Others prefer cars, or sports, or perhaps art. As an inveterate gambling man, Kimmel was a connoisseur of the favorable betting system. When Kimmel read about Thorp's blackjack system, he wrote to Thorp and offered to fund his experiment to the tune of $100,000. But first he needed to see the system in action. So when Thorp contacted him and agreed to meet, Kimmel took a car up from New York. When Kimmel arrived — introducing the two young women as his nieces — Thorp began by showing Kimmel his proofs and explaining his methodology. But Kimmel didn't care about any of that. Instead, he took a deck of cards out of his pocket and began to deal. Kimmel would believe a system worked only after he'd watched someone win with it. They played all evening, and then again the next day. Over the coming weeks, Thorp would drive down to New York regularly to play against Kimmel and an associate, Eddie Hand, who was putting up part of the money for the casino trip.

It took about a month, but at last Kimmel was convinced that Thorp's system worked — and that Thorp had what it took to use the system in a real casino. Thorp decided that $100,000 was too much and insisted on working with a smaller sum — $10,000 — because he thought gambling with too much money would attract unwanted attention. Kimmel, meanwhile, thought that Las Vegas was too high profile, and besides, too many people knew him there. So over MIT's spring break, Thorp and Kimmel, who was once again accompanied by a pair of young women, descended on Reno to test Thorp's system.

It was a resounding success. They played, moving from casino to casino, until they developed a reputation that moved faster than they could. In just over thirty man-hours of playing, Thorp, Kimmel, and Hand collectively turned their $10,000 into $21,000—and it would have been $32,000 if Kimmel hadn't insisted on continuing to play one evening after Thorp announced he was too tired to keep counting. Thorp would later tell the story—with Kimmel's name changed to Mr. X and Hand's to Mr. Y—in a book, *Beat the Dealer,* that taught readers how to use his system to take Vegas to the cleaners themselves.

Thorp developed several methods for keeping track of how the odds in blackjack change as cards are played and removed from the deck. Using these systems, Thorp was able to reliably determine when the deck was in his favor, and when it was in the house's favor. But suppose you are playing a game of blackjack, and suddenly you learn that the odds are slightly in your favor. What should you do?

It turns out that blackjack is extremely complicated. To make the problem tractable, it's better to start with a simpler scenario. Real coins come up heads and tails equally often. But it's possible to at least imagine (if not manufacture) a coin that is more likely to come up one way or the other—for now, suppose it's more likely to come up heads than tails. Now imagine you're making bets on coin flips with this weighted coin, against someone who is willing to pay even money on each flip, for as many flips as you want to play (or until you run out of money). In other words, if you bet a dollar and win the bet, your opponent gives you one dollar, and if your opponent wins, you lose one dollar. Since the coin is more likely to come up heads than tails, you would expect that over the long run money will tend to flow in one direction (yours, if you consistently bet heads) because you're going to win more than half the time. Finally, imagine that your opponent is willing to take arbitrarily large or small bets: you could bet $1, or $100, or $100,000. You have some amount of money in your pocket, and if it runs out, you're sunk. How much of it should you bet on each coin flip?

One strategy would be to try to make bets in a way that maximizes the amount of money you could stand to make. The best way to do this would be to bet everything in your pocket each time. Then, if you win, you double your money on each flip. But this strategy has a big problem: the coin being weighted means that you will *usually* win, not that you'll *always* win. And if you bet everything on each flip, you'll lose everything the first time it comes up tails. So even though you were trying to make as much money as possible, the chances that you'll end up broke are quite high (in fact, you're essentially guaranteed to go broke in the long run), with no chance to make your money back. This scenario — where your available funds run out, and you're forced to accept your losses — is known as "gambler's ruin."

There's another possibility — one that minimizes the chances of going broke. This is also a straightforward strategy: don't bet in the first place. But this option is (almost) as bad as the last one, because now you guarantee that you won't make any money, even though the coin is weighted in your favor.

The answer, then, has to be somewhere in the middle. Whenever you find yourself in a gambling situation where you have an advantage, you want to figure out a way to keep the chances of going broke to a minimum, while still capitalizing on the fact that in the long run, you're going to win most of the bets. You need to manage your money in a way that keeps you in the game long enough for the long-term benefits to kick in. But actually doing this is tricky.

Or so it seemed to Thorp when he was first trying to turn his analysis of card-counting odds into a winning strategy for the game. Fortunately for Thorp, Shannon had an answer. When Thorp mentioned the money management problem to Shannon, Shannon directed Thorp to a paper written by one of Shannon's colleagues at Bell Labs named John Kelly Jr. Kelly's work provided the essential connection between information theory and gambling — and ultimately the insights that made Thorp's investment strategies so successful.

Kelly was a pistol-loving, chain-smoking, party-going wild man from Texas. He had a PhD in physics that he originally intended to use in oil exploration, but he quickly decided that the energy industry had little appreciation of his skills, and so he moved to Bell Labs. Once he was in New Jersey, Kelly's colorful personality attracted plenty of attention in his staid suburban neighborhood. He was fond of firing plastic-filled bullets into the wall of his living room to entertain houseguests. He was an ace pilot during World War II and later earned some local notoriety by flying a plane underneath the George Washington Bridge. But despite the theatrics, Kelly was one of the most accomplished scientists at AT&T — and the most versatile. His work ran the gamut from highly theoretical questions in quantum physics, to encoding television signals, to building computers that could accurately synthesize human voices. The work he's best known for now, and that was of greatest interest to Thorp, was on applying Shannon's information theory to horseracing.

Imagine you're in Las Vegas, betting on the Belmont Stakes, a major horserace held in Elmont, New York. The big board in the off-track-betting room shows various odds: Valentine at 5 to 9, Paul Revere at 14 to 3, Epitaph at 7 to 1. These numbers mean that Valentine is expected to have a roughly 64% chance of winning, Paul Revere has an 18% chance of winning, and Epitaph has a 13% chance of winning. (These percentages are calculated by dividing the odds of each horse winning by the sum of the odds of that horse winning and losing — so, for Valentine, if your odds are 5 to 9, you divide 9 by 14.)

In the first half of the century, there was often a delay in communicating racing results between bookies. This meant that sometimes a race would be over, while people in other parts of the country continued betting on it. So if you had a particularly fast method of communication, you could in principle get the results before betting closed. By 1956, when Kelly wrote his paper, this had become quite difficult: telephones and television meant that bookies in Las Vegas would know what had happened in New York almost as soon as the people in Elmont. But suspend disbelief for a moment and imagine that you had someone in Elmont who could send you messages about the Belmont Stakes instantaneously — faster, even, than the bookies got their results.

If the messages you were receiving over your private wire service were perfectly reliable, you'd be wise to bet everything, since you're guaranteed to win. But Kelly was more interested in a slightly different case. What happens if you have someone send you correct racing results, but there's noise on the line? If the message that comes along is so garbled that you can't make out much of anything, your default guess is going to be that Valentine is going to win, since that's what the odds were to begin with and you haven't received any new information. If it's garbled but you're pretty sure you heard a *t* sound, you've gotten some information — you have good reason to think Paul Revere didn't win, since there's no *t* in his name. If pressed, you would probably guess that your contact said "Valentine," because that's the more likely message, but you can't know for sure. You wouldn't want to put all of your money on one horse, because you still have a chance of losing. But you can rule out one possibility, which gives you an advantage: you now know that the bookie thinks Valentine's and Epitaph's chances aren't as good as they really are, because the bookie is assuming Paul Revere has an 18% chance of winning. So if you make a combined bet on *both* Valentine and Epitaph in the right proportions, you're guaranteed to win one of them for a net profit. Hence even the partial information is enough to help you decide what bets to place.

Shannon's theory tells you how much credence to give a message when there's a chance that the message is being distorted by noise, or when the level of noise makes it difficult to interpret the message in the first place. So if it's difficult to decipher your racing tips, Shannon's theory provides a way of deciding how to place your bets based on the partial information you *do* receive.

Kelly worked out the solution to this problem, provided you want to maximize the long-term growth of the money you start with. As in the example above, where you could make out a *t* sound but nothing else, partial information can be sufficient to give you an advantage over a bookie who is setting odds without any information about how the race turned out. The advantage can be calculated by multiplying the payout — the number *b* when someone gives you *b*-to-1 odds — by what you believe is the true probability of winning (based on your partial information), and then subtracting the probability of losing (again, based on your partial information). To figure out how much of your starting money to bet, as a fraction of what you have, you divide your advantage by the payout. This gives the equation now called the Kelly criterion or Kelly bet size. The percentage of your money to bet on any given outcome is

$$\frac{\text{advantage}}{\text{payout}}$$

If your advantage is zero (or negative!), Kelly says not to bet at all; otherwise, bet the fraction of your wealth given by the Kelly criterion. If you always follow this rule, you will be guaranteed to outperform anyone adopting another betting strategy (such as betting it all or betting nothing). One of the most surprising things in Kelly's paper, something that feels almost mystical, is a proof of what will happen if you follow his rule in a scenario like the horse-betting story, where you have a stream of (partial) information coming in: if you always use the Kelly criterion, under certain ideal circumstances your wealth will increase at exactly the rate that information comes in along the line. Information is money.

When Shannon showed Kelly's paper to Thorp, the last piece of the blackjack puzzle fell into place. Card counting is a process by which you gain information about the deck of cards—you learn how the composition of the deck has changed with each hand. This is just what you need to calculate your advantage, as Kelly proposed. Information flows and your money grows.

As Thorp and Kimmel made their preparations for Reno, Shannon and Thorp were collaborating on Thorp's roulette plan. When he heard Thorp's ideas, Shannon was mesmerized, in large part because Thorp's roulette idea combined game theory with Shannon's real passion: machines. At the heart of the idea was a wearable computer that would perform the necessary calculations for the player.

They began testing ideas for how the actual gambling would work, assuming they could make sufficient progress on the prediction algorithm. They agreed that it would take more than one person for it to go smoothly, because one person couldn't focus sufficiently on the wheel to input the necessary data and still be prepared to bet before the ball slowed down and the croupier (roulette's equivalent of a dealer) announced that betting was closed. So they decided on a two-person scheme. One person would stand near the roulette wheel and watch carefully — ideally while doing something else, so as not to attract attention. This person would be wearing the computer, which would be a small device, about the size of a cigarette pack. The input device would be a series of switches hidden in one of the wearer's shoes. The idea was that the person watching the wheel would tap his foot when the wheel started spinning, and then again when the ball made one full rotation. This would initialize the device and synchronize it to the wheel.

Meanwhile, a second person would be sitting at the table, with an earpiece connected to the computer. Once the computer had a chance to take the initial speeds of the ball and the rotor into account, it would send a signal to the person at the table indicating how to bet. It was too difficult to predict just what number the ball would fall into, as the calculations for that level of precision were far too complicated. But roulette wheels are separated into eight regions, called octants. Each octant has four or five numbers in it, arranged in an order that would seem random to someone who didn't have the roulette wheel memorized. Thorp and Shannon discovered that in many cases, they could accurately predict which *octant* the ball would fall into, narrowing the possible outcomes from thirty-eight to four or five. The computer was designed to indicate whether there was a higher-than-normal chance that the ball would fall into a particular octant. Once the person at the table received the signal, he would quickly place bets on the appropriate numbers — using a betting system based on the Kelly criterion to decide how much to bet on each.

By the summer of 1961, the machine was ready for action. Thorp, Shannon, and their wives traveled to Las Vegas. Aside from broken wires and the night the earpiece was discovered, the experiment was a (middling) success. Unfortunately, technical difficulties prevented Thorp and Shannon from betting any substantial amounts of money, but it was clear that the device did what it was intended to do. With Shannon's help, Thorp had beaten roulette.

The trip as a whole, though, proved more stressful than it was worth. Gambling can be tense enough without the constant possibility that burly enforcers will descend on you. Meanwhile, Thorp had already received the job offer in New Mexico when the two couples made their Vegas trip. Despite their small profit, by the time they left Vegas, Thorp knew that he and Shannon wouldn't continue the roulette project. But it was just as well. With the blackjack and roulette experiences under his belt, Thorp was ready to try his hand at a new, bigger challenge: the stock market.

Thorp bought his first share of stock in 1958, before he had finished his PhD. He was living on a modest salary as an instructor at UCLA, but he had managed to cobble together a small sum to put away for the future. Over the next year, his investment dropped by half, and then slowly inched its way back up. Thorp sold, essentially breaking even after a year-long roller-coaster ride.

In 1962, flush with blackjack winnings and the proceeds from his card-counting book, he decided to try again. This time he bought silver. In the early 1960s, demand for silver was sky-high — so high that many people expected the open-market value of the silver in U.S. coins to exceed the coins' denominations, which would make quarters and silver dollars more valuable as scrap metal than as money. It seemed like a safe bet. To maximize his profits, Thorp borrowed some money from his broker, with the silver investment as collateral. Silver went up for most of the sixties, but it was very volatile; not long after Thorp bought in, the price fell temporarily, but sharply, and the broker decided he wanted his money back. When Thorp couldn't come up with the cash, the broker sold Thorp's silver, at a loss of about $6,000 to Thorp. It was devastating — over half the annual salary for an assistant professor in 1962.

After this second setback, Thorp decided to get serious. After all, he was a world-renowned expert in the mathematics of gambling. And the stock market wasn't so different from a casino game or a horserace: you make bets, based on some partial information about the future, and if things go your way, you get a payout. You can even think of market prices as reflections of the "house" odds, meaning that if you can get access to even partial relevant information, you can compare market odds and true odds to determine whether you have an advantage, just as in blackjack.

All Thorp needed was to figure out a way to get information. Thorp began his careful study of markets in the summer of 1964, by reading *The Random Character of Stock Prices* — the collection of essays that featured papers by Bachelier, Osborne, and Mandelbrot. Thorp was soon convinced by Osborne and the other authors in the collection who argued that when you look at the detailed statistics, stock prices really *do* behave randomly — because, as Bachelier and Osborne both argued, all available information was already incorporated into the price of a stock at any given moment. By the end of the summer, Thorp was stymied. If Osborne was right, Thorp didn't see a way to gain an advantage over the market.

With a full teaching load for the 1964–65 academic year, Thorp had little time for anything else. He put his market studies aside, planning to return to the project the following summer. In the meantime, things in New Mexico took a turn. A growing faction of mathematicians working in a different field had taken control of the department, prompting him to look for other jobs. He learned that the University of California was preparing to open a new campus, about fifty miles south of Los Angeles, in the middle of Orange County. He applied for, and received, a job at the new University of California, Irvine. It looked as if work on the stock market would have to be deferred further, since he now had another major move to plan and a new department to settle into.

Still, he remained interested in the project, and at some point during the year, while scanning advertisements in investment magazines, Thorp came across a publication called the *RHM Warrant Survey*. Warrants are a kind of stock option, offered directly by the company whose stock is being optioned. Like an ordinary call option, they give the holder the right to purchase a stock at a fixed price, before a fixed expiration date. Throughout the middle of the twentieth century, options weren't traded widely in the United States. Warrants were the closest thing to an option available. *RHM* claimed that trading warrants was a possible source of untold wealth — if you understood them. Implicit was that most people didn't know what to do with warrants. This was just the kind of thing Thorp was looking for, and so he decided to subscribe. But he didn't have much time to look at the documents that began arriving.

As the spring semester came to an end in New Mexico, Thorp found himself with a few weeks to spare before his move to California. He began to riffle through the *RHM* documents. The writers at *RHM* apparently thought of warrants as a kind of lottery ticket. They were cheap to buy, usually worthless, but occasionally you could strike it rich if a stock started trading well above the warrant's exercise price.

Where *RHM,* and most other investors, saw a lottery ticket, Thorp saw a bet. A warrant is a bet on how a stock will perform over a fixed period. The price of the warrant, meanwhile, is a reflection of the market's determination of how likely the buyer is to win the bet. It also reflects the payout, since your net profit if the warrant does become valuable is determined by how much you had to pay for the warrant in the first place. But Thorp had just spent an entire summer reading about how stock prices are *random.* He pulled out a piece of paper and began to calculate. His reasoning followed Bachelier's thesis closely, except that he assumed prices were log-normally distributed, à la Osborne. He quickly arrived at an equation that told him how much a warrant should really be worth.

This was valuable, if not trailblazing. But Thorp had an ace up his sleeve, something Bachelier and Osborne never imagined. With five years of gambling experience, Thorp realized that calculating a "true" price for a warrant is a lot like calculating the "true" odds on a horserace. In other words, the theoretical relationship that Thorp discovered between stock prices and warrant prices gave him a way to extract information from the market — information that gave him an edge, not in the stock market directly, but in the associated *warrant* market. This partial information was just what Thorp needed to implement the Kelly system for maximizing long-term profits.

Thorp was energized by this work on warrants. It seemed to him that he had finally found the perfect way to use his gambling experience to profit from the world's biggest casino. But there was a problem. When he finished his calculations and plugged some numbers into a computer (Thorp wasn't able to solve the equations he set up explicitly, but he was able to come up with a way to use a computer to do the final calculations for him), he discovered that there was no advantage to buying warrants. In other words, you couldn't go out and buy warrants and expect to make a profit — according to the Kelly betting system, you should invest nothing! The reason for this wasn't that warrants were all trading at exactly what they were worth; rather, they were trading at much too high a price. The dirt-cheap lottery tickets that *RHM Warrant Survey* was advertising were actually much, much too expensive.

If you think of investing as a kind of gamble, buying a stock represents a bet that the stock price will go up. Selling a stock, meanwhile, is a bet that the stock will go down. Thorp, like Bachelier before him, realized that the "true" price of a stock (or option) corresponds to the price at which the odds of the buyer winning are the same as the odds of the seller winning. But with traditional trades, there's an asymmetry. You can virtually always buy a stock; but you can sell a stock only if you already own it. So you can bet *against* a stock only if you've already chosen to bet *for* it. This is similar to a casino: it would be highly desirable, in roulette, say, to bet *against* a number. This, after all, is what the house does, and the house ultimately has the long-term advantage. But it isn't possible. No casino will let you bet that your blackjack hand will lose.

In investing, however, there *is* that possibility. If you want to sell a stock you don't already own, all you need to do is find someone who *does* own the stock but doesn't want to sell it, and who is willing to let you borrow the shares for a while. Then you sell the borrowed shares, with the expectation that at some later time you will buy the same number of shares back and return them to their original owner. This way, if the price goes down after you sell, you see a profit, since you can buy the shares back at the lower price. Whoever loaned you the shares, meanwhile, is no worse off than if he had simply held on to them. The origins of this investment practice, known as short selling, are obscure, but it is at least three hundred years old. We know this because it was banned in England in the seventeenth century.

Today, short selling is perfectly standard. But in the 1960s — indeed, for much of the practice's history — it was viewed as dangerous at best, and perhaps even depraved or unpatriotic. The short seller was perceived as a blatant speculator, gambling on market moves rather than investing capital to spur growth. Worse, he had the nerve to take a financial interest in bad news. This struck many investors as déclassé. Views on short selling changed in the 1970s and 1980s, in part because of Thorp's and others' work, and in part because of the rise of the Chicago School of economics. As those economists argued at the time, short selling may seem crude, but it serves a crucial social good: it helps keep markets efficient. If the only people who can sell a stock are the ones who already own it, people who have information that could be *bad* for the company often don't have any way of affecting market prices. This would mean that information could be available that *isn't* reflected in the stock price, because the people who have access to the information aren't able to participate in the market. Short selling prevents this situation.

Whatever the social impact, short selling does have real risks attached. When you buy a stock (sometimes called taking a "long" position, in contrast to the "short" position that short sellers take), you know how much money you stand to lose. Stockholders aren't responsible for a corporation's debts, so if you put $1,000 into AT&T, and AT&T goes under, you lose at most $1,000. But stocks can go *up* arbitrarily high. So if you make a short sale, there's no telling how much money you stand to lose. If you sell $1,000 worth of AT&T short, when it comes time to buy the shares back to repay the person you borrowed them from, you might need to come up with a lot more money than you originally received in the sale in order to get the shares back.

Still, Thorp was able to find a broker who was willing to execute the required trades. This solved one problem, of figuring out how to apply Kelly's results in the first place. But even if Thorp could ignore the social stigma of short selling — and he could — the real dangers of unlimited losses remained. Here, though, Thorp had one of his most creative insights. His analysis of warrant pricing gave him a way of relating warrant prices to stock prices. Using this relationship, he realized that if you sell warrants short, but at the same time you buy some shares of the underlying stock, you can protect yourself against the warrant increasing in value — because if the warrant increases in value, according to Thorp's calculations the stock price should *also* increase, limiting your losses on the warrant. Thorp discovered that if you pick the right mix of warrants and stocks, you can guarantee a profit unless the stock price moves dramatically.

This strategy is now called delta hedging, and it has spawned other strategies involving other "convertible" securities (securities that, like options, can be exchanged for another security, such as certain bonds or preferred shares of stock that can be converted to shares of common stock). Using such strategies, Thorp was able to consistently make 20% per year . . . for about forty-five years. He's still doing it — indeed, 2008 was one of his worst years ever, and he made 18%. In 1967, he wrote a book, called *Beat the Market,* with a colleague at UC Irvine who had worked on similar ideas.

Beat the Market was too unusual, too different from then-current practices, to change Wall Street overnight. Many traders simply ignored it; most who read it didn't understand it, or missed its importance. But one reader, a stockbroker named Jay Regan, saw Thorp's genius. He wrote to Thorp and proposed that they enter a partnership to create a "hedge fund." (The term *hedge fund,* originally "hedged" fund, was already twenty years old when Thorp and Regan first met, but nowadays so many hedge funds are based on ideas related to Thorp's delta hedging strategy that the name might as well have originated with Thorp and Regan.) Regan would take care of the tasks that Thorp hated: he would promote the fund, find and manage clients, interface with brokers, execute the trades. Thorp would just be responsible for identifying the trades and working out the mix of stocks and convertibles to buy and sell. Thorp wouldn't even have to leave the West Coast: Regan was happy to run the business end of things from New Jersey, while Thorp stayed in Newport Beach, California, building a team of mathematicians, physicists, and computer scientists to identify favorable trades. The deal seemed too good to be true. Thorp quickly agreed.

The company that Thorp and Regan created was initially called Convertible Hedge Associates, though in 1974 they changed the name to Princeton-Newport Partners. Success came quickly. In its first full year, their investors made just over 13% each on their investments, after fees — while the market returned only 3.22%. They also had some impressive early admirers. One of their earliest investors, Ralph Gerard, the dean of UC Irvine's graduate school — in a sense, Thorp's boss — had inherited a fortune. He was looking to invest with a new fund, because his old money manager was moving on to other projects. Thorp was close to home, but before Gerard would invest with the new partnership, he wanted his old money manager, a trusted friend, to take a careful look at Thorp. Thorp agreed to the meeting, and one evening he and Vivian drove a few miles down the Pacific Coast Highway, to Laguna Beach, where the old money manager lived. The plan was to play bridge and chat casually, so that the old money manager could size Thorp up.

Thorp learned that his host was leaving the money management business to focus on a new venture — an old manufacturing and textiles company that he was hoping to rebuild. He'd made his first million managing other people's money, and now it was time to put his own money to work. But mostly, Thorp and his host discussed probability theory. While they were playing, the host mentioned a kind of trick dice, called nontransitive dice. Nontransitive dice are a set of three dice with different numbers on each side. They have the unusual property that if you roll dice 1 and 2 at the same time, die 2 is favored; if you roll dice 2 and 3 at the same time, die 3 is favored; but if you roll dice 1 and 3 at the same time, die 1 is favored. Thorp, always a fan of games and the probabilities associated with them, had long been interested in nontransitive dice. From that point on, the two were fast friends. On the ride back to Newport Beach, Thorp told Vivian that he expected their host to someday be the richest man in the world. In 2008, his prediction came true. The old money manager's name was Warren Buffett. And at his recommendation, Gerard invested with Thorp's company.

Princeton-Newport Partners quickly became one of the most successful hedge funds on Wall Street. But all good things must end. And Princeton-Newport's demise was particularly dramatic. On December 17, 1987, about fifty FBI, ATF, and Treasury Department agents pulled up in front of the firm's Princeton office. The agents stormed into the building, looking for records and audiotapes regarding a series of trades the firm had made with the soon-to-be-indicted junk bond dealer Michael Milken. A former Princeton-Newport employee named William Hale had testified to a grand jury that Regan and Milken were engaged in a tax dodge known as stock parking. One downside to delta hedging and related strategies is that profits from short-term and long-term positions are taxed differently. So when you buy and sell at the same time, profits and losses that otherwise would cancel each other out *don't* cancel from a tax perspective. Regan was trying to avoid additional taxes by concealing who actually owned the long-term positions, "parking" the stocks at Milken's firm. Parked stocks were officially sold to Milken, with an unofficial agreement that Regan could buy them back for a predetermined price, irrespective of what had happened in the market in the meantime. Though hardly nefarious, stock parking was illegal, and Rudy Giuliani, who was prosecuting the case, hoped that by applying pressure on the Princeton-Newport side, he could dig up additional evidence against Milken.

By all accounts, Thorp was completely in the dark. He didn't know that the East Coast side of the firm was doing anything illegal until the scandal broke in the news. He was never accused of, let alone charged with, any crime. And by the time he got wind of the raid, Regan already had a lawyer and refused to talk to his partner. The firm hobbled along for another year, but the legal proceedings had ruined its reputation. In 1989, Princeton-Newport Partners closed. Over the course of twenty years, it had average returns of 19% (over 15% after fees) — an unprecedented performance.

After Princeton-Newport closed, Thorp took some time off before regrouping to form Edward O. Thorp Associates, his own money management firm. Though he has long since given up managing other people's money professionally, he still runs the fund today using his own capital. Meanwhile, hundreds of quant hedge funds have opened (and closed), trying to reproduce Princeton-Newport's success. As the *Wall Street Journal* put it in 1974, Thorp had ushered in a "switch in money management" to quantitative, computer-driven methods. It's amazing what a little information theory can do.

CHAPTER 5

Physics Hits the Street

IN FEBRUARY 1961, Fischer Black's PhD advisor, Anthony Oettinger, wrote to Harvard's Committee on Higher Degrees: "I have reason to be concerned about [Black's] intellectual discipline so that, while recognizing his ability and his desire for independence, I am concerned lest he lapse into dilettantism." Two months later, Oettinger chaired an oral exam designed to determine whether Black was prepared to advance to the dissertation stage of his doctorate. Black passed — but with an explicit requirement that he produce "a coherent, lucid thesis outline" by January 1962. Within a week, Black was in jail for participating in student riots in Harvard Square, and when a Harvard dean went to bail him out, Black was unrepentant. He railed against police authority, against Harvard's authority, and against his advisor. January 1962 came and went, and Black had done no more work toward his thesis. He was informed that he could not return to Harvard.

Today, Fischer Black is one of the most famous figures in the history of finance. His most important contribution, the Black-Scholes (sometimes Black-Scholes-Merton) model of options pricing, remains the standard by which all other derivatives models are measured. In

1997, Black's collaborators, Myron Scholes and Robert Merton, were awarded the Nobel Prize in economics for the Black-Scholes model. Black had died in 1995 and so was ineligible for the prize (the Nobel is never awarded posthumously), but in a rare nod, the Nobel committee explicitly acknowledged Black's contribution in its announcement of the award. Every two years, the American Finance Association awards the Fischer Black Prize — one of the most prestigious awards in academic finance — to an individual under forty whose body of work "best exemplifies the Fischer Black hallmark of developing original research that is relevant to finance practice." MIT's Sloan School of Management endowed a chair in financial economics in Black's honor. And the list goes on.

In the broad history of physics in finance, Black is perhaps best seen as a transitional figure. He was trained as a physicist but was never successful as one — in large part because he was too wide-ranging and unfocused. Though he was more successful as a financial economist, his career was fleeting, as he quickly became bored with the projects that made him famous and turned to new ideas that were met with much more skepticism. Yet these very qualities — the qualities that Oettinger worried would lead Black to dilettantism — were what allowed Black to bring about a marriage long in the waiting. He was enough of a physicist to understand and develop the insights of people like Bachelier and Osborne, and yet he was enough of an economist to express his discoveries in a language economists could understand. In these ways he was like Samuelson, though he was never as intellectually distinguished. But unlike Samuelson, Black was able to communicate to investors and Wall Street bankers how the new ideas coming out of physics could be used in practice. Thorp was the first person to figure out how to use Bachelier's and Osborne's random walk hypothesis to make a profit, but he did so outside of the establishment, through Princeton-Newport Partners. Black, on the other hand, was the person who made quantitative finance, with its deep roots in physics, an essential part of investment banking. Black took physics to the Street.

Black first arrived at Harvard in 1955, at age seventeen. If anyone asked why he applied to Harvard and nowhere else, he would say it was because he liked to sing and Harvard had a great glee club. From the very beginning, he was determined to chart his own course through academia. He refused to do the work he was assigned and instead wrote papers on topics *he* decided were interesting. After a few semesters of introductory courses, he decided to enroll in graduate classes. He picked an interdisciplinary major called "social relations," which combined several social science disciplines, and then promptly began conducting experiments with himself as the subject. For instance, he would modify his sleep schedule, alternating between four hours awake and four hours asleep, all while taking careful and copious notes on how his body reacted. He began taking drugs, including hallucinogens, and tracking the effects. Most of his friends were graduate students.

Come junior year, however, he started having second thoughts about his choice of major. Social relations was interesting, but Black wanted a career in research. Like Osborne and Thorp, Black was a natural-born scientist, constantly experimenting and coming up with theories to test, and he just didn't see how social relations could get him the kind of job he wanted. So he turned toward the hard sciences, flirting with chemistry and biology before finally settling on physics. He wanted to do fundamental, theoretical work, and so the next year he applied to graduate school, once again only to Harvard, to do a PhD in theoretical physics. He won a prestigious National Science Foundation graduate student fellowship and Harvard admitted him. In the fall of 1959, Black started graduate school as a physicist.

But by the end of his first year, his attention had begun to stray again. He took only one physics course, filling his first year instead with electrical engineering, philosophy, and mathematics. He was a little interested in everything, but not enough interested in anything to stay focused for long. After just a few weeks, he switched departments, to study applied mathematics instead of physics; then, come spring semester, he was devoting all of his time to an artificial intelligence course at MIT, taught by AI pioneer Marvin Minsky; by fall 1960, he was back to the social sciences, taking two courses in psychology.

It would be wrong to say that Black did poorly in school. But his tack was certainly unconventional. On the one hand, he barely passed some courses — including the one physics course he enrolled in. During his second year he failed a psychology course because it emphasized "behavioralist" methods, while Black saw himself as aligned with the newer, more fashionable "cognitivist" school. But he was certainly one of the best minds at Harvard. In an open competition during his first year, he successfully solved a challenge problem offered by one of his mathematics professors, which earned him an endowed scholarship for the following year. And so his abilities were never really in doubt. It is nonetheless easy to see Oettinger's worry: two years into graduate school, Black was no closer to settling on a major than he was as an undergraduate. If anything, the rate at which he swung from discipline to discipline was accelerating. As Black saw it, he was simply curious, and he wasn't going to be pinned down by some stodgy old school's rules about what constituted appropriate academic work — even if it meant leaving Harvard.

Ultimately, Black did earn a PhD in applied mathematics. But he took the scenic route. When Harvard asked him to leave, he found a job at Bolt, Beranek and Newman (BBN), a Cambridge-based high-tech consulting firm. BBN hired Black because of his computer skills, and most of his time there was spent working on computerized data retrieval systems for a project commissioned by the Council on Library Resources. As part of this project, Black wrote a program that used formal logic to try to answer simple questions. The program would take an input such as "What is the capital of Romania?" and try to deduce an answer based on a list of facts it had stored in a database. A major part of this project was devoted to simply parsing the question, trying to determine what the questioner was even after. Black's work represented an important early contribution to the field known as computational linguistics, in which people try to figure out how to make computers understand and produce natural language.

Word spread quickly around Cambridge of Black's work at BBN. In the spring of 1963, Minsky heard about Black's question-answering program. He was sufficiently impressed — and sufficiently influential — that he negotiated readmission to Harvard on Black's behalf. Minsky took responsibility for Black's work, with a professor at Harvard named Patrick Fischer serving as the official advisor. Over the next year, Black turned his consulting project into a dissertation on deductive question-answering systems, which he successfully defended in June 1964.

But by this time, Black had had enough of academia — at least for a while. He had settled on a project long enough to write a dissertation, but this hardly signaled a lifelong devotion to artificial intelligence. He thought about becoming a writer, working on popular nonfiction projects. Or maybe he would go into the computer business. He considered applying for a postdoctoral fellowship to stay at Harvard and work on the interface between technology and society — a new subject, spurred by new postwar technologies. But ultimately nothing panned out, and so after graduating, Black returned to consulting. At least there, he could work on many different projects, and he had already discovered that solving concrete problems appealed to him.

Instead of returning to BBN, however, Black took a job with another local firm, called Arthur D. Little, Inc. (ADL), in the Operations Research Division. At first Black worked primarily on computer problems. For instance, MetLife had a state-of-the art computer, but the company still felt as if its computation needs weren't being met. MetLife hired ADL to see if a second computer was needed. Black, collaborating with two others at ADL, discovered that the problem wasn't the computer, which was working at only half capacity, but rather the way in which the computer stored data: instead of using thirty available drives, it used only eight drives in everyday tasks. So Black and his team worked out an optimization scheme for using all of the available drives.

Black worked at ADL for about five years. The experience changed his life. When he arrived, he was an operations research and computer science guy. He had unusually broad interests, but there's no evidence to suggest that finance was among them. When he left in 1969, he had already laid the foundation for the Black-Scholes model. He was recognized, at least in some circles, as an exciting, if radical, up-and-coming financial economist. Wells Fargo immediately hired him to develop a trading strategy.

This transformation began shortly after Black arrived at ADL, where he encountered a slightly older member of the operations research section named Jack Treynor. Treynor had gone to Haverford College intending to major in physics but decided that the department wasn't very good, and so he switched to mathematics. After college he went to Harvard Business School and then joined ADL in 1956, a decade before Black would arrive. Treynor and Black didn't overlap at ADL for long: in 1966, Treynor was wooed away by Merrill Lynch. But the two practically minded mathematicians made fast friends. Black liked Treynor's way of thinking and quickly became interested in his work, primarily on risk management, hedge fund performance, and asset pricing. Although Treynor didn't have a formal background in financial theory either, his business school background had exposed him to a set of problems that he was well suited to work on, and so much of his work at ADL involved financial institutions. Meanwhile, he worked on more theoretical research projects on the side, often motivated by the kinds of problems ADL clients encountered.

By the time Black arrived at ADL, Treynor had already developed a new way of understanding the relationship between risk, probability, and expected value, now known as the Capital Asset Pricing Model (CAPM). The basic idea underlying CAPM was that it should be possible to assign a price to risk. Risk, in this context, means uncertainty, or volatility. Certain kinds of assets — U.S. Treasury bonds, for instance — are essentially risk-free. Nonetheless, they yield a certain rate of return, so that if you invest in Treasury bonds, you are guaranteed to make money at a fixed rate. Most investments, however, are inherently risky. Treynor realized that it would be crazy to put your money into one of these risky investments, unless you could expect the risky investments to have a higher rate of return, at least on average, than the risk-free rate. Treynor called this additional return a risk premium because it represented the additional income an investor would demand before buying a risky asset. CAPM was a model that allowed you to link risk and return, via a cost-benefit analysis of risk premiums.

When Black learned about CAPM, he was immediately hooked. He found the simple relationship between uncertainty and profit deeply compelling. CAPM was a big-picture theory. It described the role of risk in making rational choices in a very abstract way. Later in his career, Black would point to one feature of CAPM in particular that he was drawn to: it was (in his words) an equilibrium theory. "Equilibrium was the concept that attracted me to finance and economics," Black wrote in 1987. CAPM was an equilibrium theory because it described economic value as the natural balance between risk and reward. The idea that the world was in a constantly evolving equilibrium would have appealed to Black's sensibilities as a physicist: in physics, one often finds that complicated systems tend toward states that are stable under small changes. These states are called equilibrium states because they, too, represent a kind of balance between different influences.

Black set out to learn everything that Treynor knew about finance, so that when Treynor left ADL, just a year after he and Black first met, Black was the natural person to take Treynor's place on ADL's financial consulting team — and further perfect Treynor's model. CAPM would form the foundation for virtually all of the work Black would go on to do.

If Jack Treynor initiated Black's transformation into a financial economist, Myron Scholes brought it to fruition. Scholes arrived in Cambridge in September 1968, a fresh doctorate from the University of Chicago in hand. A fellow graduate student in Chicago, Michael Jensen, had recommended that Scholes look up Black — an "interesting fellow," in Jensen's estimation. Scholes called soon after arriving in Cambridge. Both men were young: Scholes had just turned twenty-seven, and Black was thirty. Neither was particularly accomplished, though Scholes's recent appointment as an assistant professor at MIT was a promising sign. They met over lunch in the drab, institutional cafeteria at ADL's Acorn Park campus. One rarely imagines history unfolding over a cafeteria meal shared by undistinguished men. And yet, that first meeting between Black and Scholes was the start of a friendship that would change financial markets forever.

Black and Scholes were polar opposites. Black was quiet, even shy. Scholes was outgoing and brash. Black was interested in applied work, but he had a theoretical, abstract mind. Scholes, meanwhile, had just written a heavily empirical thesis, analyzing piles of data to test the efficient market hypothesis, which by this point had been elevated to a central principle of neoclassical economics. It is difficult to imagine how that first conversation could have gone. And yet, something clearly clicked. The two men met again, and then again. Soon they had laid the foundation for a lifelong friendship and intellectual partnership. Scholes invited Black to participate in the weekly MIT finance workshop, which was Black's first opportunity to fully engage with finance academics. Soon after, Wells Fargo approached Scholes with an offer for a consulting arrangement, to help the bank implement some of the new ideas in finance, like CAPM, that were just bubbling to the surface in academia. Scholes felt that he didn't have enough time to do the work himself, but he knew someone who would be perfect for the job. Black quickly agreed, and in March 1969, some six months after that first meeting in the ADL cantina, Black quit his job at ADL and went off on his own. He started a new consulting firm, called Associates in Finance, with Wells Fargo as his principal client. He and Scholes were tapped to help Wells Fargo create a new, state-of-the-art investment strategy.

It was around this time that Black began thinking about ways to extend CAPM to different kinds of assets and different kinds of portfolios. For instance, he tried to apply CAPM to the question of how to apportion one's investments over time. Should you change your risk exposure as you get older, as some people were suggesting? Black decided the answer was no: just as you want to diversify over different stocks at a given time, you also want to diversify over different times, to minimize the impact that any particular stretch of bad luck might have. The question of how to value options using CAPM was just one of the many such problems that Black was working on at this time. And as early as the summer of 1969, Black had already made progress, by deriving the fundamental relationship that would ultimately give rise to the Black-Scholes equation.

The essential insight was that at any given instant, it is always possible to create a portfolio consisting of a stock and an option on that stock that would be perfectly risk-free. If this sounds familiar, it's because the idea is very similar to the one at the heart of Thorp's delta hedging strategy: he, too, realized that if the prices of options and their underlying assets are related, you could combine options and stocks to control risk. The difference was that Thorp's delta hedge strategy aimed to guarantee a profit, provided that the underlying stock's price didn't change too dramatically. This approach controlled risk, but it didn't eliminate it altogether. (Indeed, if CAPM-style reasoning is correct, you shouldn't be able to both eliminate risk and still make a substantial profit.) Black's approach was to find a portfolio consisting of stocks and options that was risk-free, and then argue by CAPM reasoning that this portfolio should be expected to earn the risk-free rate of return. Black's strategy of building a risk-free asset from stocks and options is now called dynamic hedging.

Black had read Cootner's collection of essays on the randomness of markets, and so he was familiar with Bachelier's and Osborne's work on the random walk hypothesis. This gave him a way to model how the underlying stock prices changed over time—which in turn gave him a way to understand how options prices must change over time, given the link he had discovered between options prices and stock prices. Once Black had found this fundamental relationship between the price of a stock, the price of an option on that stock, and the risk-free interest rate, it was just a few steps of algebra for him to derive an equation for the value of the option, by relating the risk premium on the stock to the risk premium on the option. But now he was stuck. The equation he derived was a complicated differential equation—an equation relating the instantaneous rate of change of the price of the option to the instantaneous rate of change of the price of the stock—and Black, despite his background in physics and mathematics, didn't know enough math to solve it.

After struggling for several months, Black gave up. He didn't tell anyone about the options problem, or his partial solution, until later in 1969, when Scholes mentioned that one of his master's students at MIT was interested in options pricing. Scholes began to speculate about whether CAPM could be used to solve the problem—at which point Black opened his desk drawer and pulled out a sheet of paper with the crucial differential equation written on it, and from then on the two men worked on the problem together. They had solved it by summer 1970, and the Black-Scholes equation for the price of an option made its debut in July, at a conference that Scholes organized at MIT, sponsored by Wells Fargo. In the meantime, a new colleague of Scholes's at MIT, Robert Merton (himself an engineer by training, though he went on to earn a PhD in economics), had rederived the same differential equation and the same solution from an entirely different starting point. With two different approaches giving the same answer, Black, Scholes, and Merton were convinced they were on to something big.

Black and Scholes submitted their paper to the *Journal of Political Economy*, one of the most important publications in the field, soon after they had solved the problem. The paper was promptly rejected, without so much as a note of explanation (suggesting it wasn't even seriously considered). So they tried again, this time with *Review of Economics and Statistics*. Again, they received a rapid rejection with no articulation of what was wrong with the article. Merton, meanwhile, held off on sending his alternative approach to journals, so that Black and Scholes could receive appropriate credit for their discovery.

Despite the early setbacks, however, Black and Scholes were not destined to labor in obscurity. Powerful forces in academia, in finance, and in politics were aligning in their favor. And some of the then-reigning academic gods were ready to intervene. After the second rejection, University of Chicago professors Eugene Fama and Merton Miller, two of the most influential economists at the time and leaders of the then-nascent Chicago School of economics, successfully urged the *Journal of Political Economy* to reconsider, and in August 1971 the article was accepted for publication, pending revisions.

In the meantime, Fischer Black had attracted attention at the University of Chicago. Economists there were familiar with his work with Scholes, both on options and at Wells Fargo; they'd seen him in action at the Wells Fargo conference. A few years earlier, in 1967, Black had traveled to Chicago with Treynor to present some of their collaborative work to the graybeards. Chicago economists didn't need fancy journals to vet young academics: they knew talent when they saw it, and Black certainly had talent. And so in May 1971, they offered Black a job. At this point, Black had already been out of graduate school for seven years, yet he had only four publications, just two of which were in finance. He had a PhD, but in an unrelated field. None of this mattered. Chicago wanted him.

Chicago wasn't working on a hunch that Black's work would become important. The faculty there had some inside information: options were about to become a *really* big deal—and a formula that allowed investors to price them would prove essential. Two major changes to U.S. and international policy were in the works, both centered in Chicago, that would soon revolutionize the derivatives industry. Having someone like Black on one's team could only help.

The first major change took place on October 14, 1971, just a few weeks after Black arrived in Chicago. The Securities and Exchange Commission (SEC) gave the go-ahead to the Chicago Board Options Exchange (CBOE), the first open, dedicated options market in United States history. Options had been around for hundreds of years, and they had been traded in the United States, often in the guise of warrants, since at least the middle of the nineteenth century. But they had never been traded on an open market before. Economists in Chicago had been agitating for the SEC to remove barriers to an open options exchange for years, until finally they convinced the Chicago Board of Trade (CBOT) to convene a committee to consider the possibility, in 1969. The head of that committee was James Lorie, a faculty member at the University of Chicago business school; later, Lorie and Merton Miller were essential in writing the report on the public impact of an options exchange that would become a major part of the CBOT's proposal to the SEC in March 1971.

The CBOE and the Black-Scholes paper were greenlit within months of each other; two years later, the CBOE opened for trading, just a month before the Black-Scholes article would appear in print. On the first day of trading, nine hundred options were traded on sixteen underlying stocks. But volume grew at an astonishing rate: well over a million options were traded in 1973 alone, and by October 1974, the exchange began seeing single days in which as many as forty thousand options were traded, with regular volume above thirty thousand. Within a decade, this number would reach half a million. And competition from other exchanges popped up quickly: first the American

Stock Exchange announced it would begin trading options, followed quickly by the Philadelphia and Pacific stock exchanges. In January 1977, the European Options Exchange was established in Amsterdam, modeled on the CBOE. Options trading was suddenly a big business, and, at least at first, investors were anxious to learn as much as they could about the new instruments. Black, Scholes, and Merton quickly became household names, at least in finance.

The second fortuitous policy change, as far as Black's career was concerned, occurred almost simultaneously with the creation of the CBOE, though its impact on Black was slower. Once again, Chicago's influential economists, and especially the famous monetarist Milton Friedman, were behind the initiative. In 1968, when Nixon was elected president, Friedman wrote him a letter urging him to abandon the so-called Bretton Woods system. Bretton Woods, named for the town in New Hampshire where the system was devised in July 1944, was the international monetary agreement put in place at the end of World War II. The Bretton Woods conference led to the creation of the International Monetary Fund (IMF) and the International Bank for Reconstruction and Development (now part of the World Bank). More important for our story was the fact that under the Bretton Woods system, major world currencies were valued at fixed exchange rates, based on the value of the U.S. dollar (and ultimately on gold, because the dollar was freely exchangeable for gold, at least for foreign governments). Changes in these exchange rates were infrequent, involving a long diplomatic process.

By 1968, however, when Friedman wrote to Nixon, the Bretton Woods system was beginning to show cracks. The main problem was that there simply wasn't enough gold in the world to back the explosion in postwar international trade. While the United States held most of the world's gold supply, gold continued to be traded on the open market, where its price could fluctuate. As long as the United States and its allies could keep open-market gold prices in line with the Bretton Woods price, there was no problem. But if the price of gold on the open market rose too high, as it naturally would with growing demand and a limited supply, there would be a risk of a run on the dollar (in the sense of a rush to convert dollars to gold), as foreign governments sought to settle their own debts by buying U.S. gold and selling it for a profit on the open market — in which case the system would simply collapse. Such a rush in fact occurred in late 1967, which was the impetus for Friedman to write his letter. But for a thinker like Friedman, the Bretton Woods system was ill conceived from the start: it was hopeless for governments to try to set exchange rates at all. Exchange rates, like anything else, should be determined freely in an open market.

Nixon didn't listen to Friedman at first, but by 1971, with increased spending in Vietnam accelerating the accumulation of U.S. debt, he saw the writing on the wall. First West Germany and Japan pulled out of the Bretton Woods agreement and announced their currencies would no longer maintain parity with the dollar. Then, rather than wait for the world economy to collapse, Nixon administered the *coup de grâce* to the Bretton Woods system by ending the convertibility of U.S. dollars to gold. Over the next years, the fixed exchange rates gave way to floating rates, creating a system whereby the relative prices of currencies were determined on the open market.

Meanwhile in Chicago, Leo Melamed, the chairman of the Chicago Mercantile Exchange (CME), another futures exchange that had spun off from the CBOT in the early twentieth century, saw that global fiscal policy was in flux. Following a hint from Friedman, Melamed launched a new exchange of his own in May 1972, called the International Monetary Market (IMM), for trading futures contracts in foreign currency. As long as the Bretton Woods system was in place, trading currency futures wasn't very interesting because currency values could change only through a laborious and public process. But once the exchange rate was allowed to float and be determined by open-market trading, futures markets became essential. Most important was that companies, and especially banks, could use currency futures to protect themselves against unexpected changes in currency values. Suppose that a company in the United States contracts with a company in the United Kingdom to send a shipment of cowboy boots in exchange for payment in pounds on delivery. The agreement is made at a particular time, but the payment won't come in until the cowboy boots hit Britain. And in the meantime, the pound could change in value, so that the U.S. company's profits (in dollars) would be lower than they were when the contract was made. To protect against such changes, the U.S. company could sell a futures contract for the amount it plans to receive when the shipment arrives, effectively eliminating the risk that the currency might change unexpectedly.

What does the IMM have to do with Black and Scholes's options pricing formula? At first glance nothing — but within a few years, futures trading at the IMM had expanded to include new derivatives based on currencies, including options. Because currency risk is an important part of any international transaction, currency derivatives very rapidly became essential to the international economy. And once again, as at the CBOE, the Black-Scholes model became an integral part of everyday trading life. Even more importantly, Black and Scholes pointed a way forward for modeling *other* derivatives contracts, too, which rapidly grew at the IMM as businesses sought new ways of protecting themselves against currency risk. Between the IMM and the CBOE, Black and Scholes found a world that was perfectly poised to take advantage of their new ideas.

The options pricing formula that Black, Scholes, and Merton discovered was equivalent to the method that Thorp had worked out in 1965 for pricing warrants — though Thorp used a computer program to calculate options prices, rather than derive the explicit equation that bears Black's, Scholes's, and Merton's names. But the underlying arguments were different. Thorp's reasoning followed Bachelier's: he argued that a fair price for an option should be the price at which the option could be interpreted as a fair bet. From here, Thorp worked out what the price of an option should be, assuming that stock prices satisfy the log-normal distribution Osborne described. Once he had a way of calculating the "true" price of an option, Thorp went on to work out the proportions of stocks and options necessary to execute the delta hedging strategy.

Black and Scholes, meanwhile, worked in the opposite direction. They started with a hedging strategy, by observing that it should always be possible to construct a risk-free portfolio from a combination of stocks and options. They then applied CAPM to say what the rate of return on this portfolio should be — that is, the risk-free rate — and worked backward to figure out how options prices would have to depend on stock prices in order to realize this risk-free return.

The distinction may seem inconsequential — after all, the two arguments are different paths to the same model of options prices. But in practice, it was crucial. The reason is that dynamic hedging, the basic idea behind the Black-Scholes approach, gave investment banks the tool they needed to *manufacture* options. Suppose you are a bank and you would like to start selling options to your clients. This amounts to selling your clients the right to buy or sell a given stock at a predetermined price. Ideally, you don't want to make a risky bet yourself — your profits are going to come from the commissions you will earn on the sales, not on the proceeds of speculation. In effect this means that when a bank sells an option, it wants to find a way to counterbalance the possibility that the underlying stock will become valuable, without losing money if the option *doesn't* become valuable. Black and Scholes's dynamic hedging strategy gave banks a way to do exactly this: using the Black-Scholes approach, banks could sell options and buy other assets in such a way that (at least theoretically) they didn't carry any risk. This turned options into a kind of product, something that banks could construct and sell.

Black stayed in Chicago until 1975, when MIT wooed him back to Cambridge. For a few years, academia seemed like the perfect fit for Black. He could work on whatever he liked, and at least in the early heyday of exchange-based options trading, it seemed he could do no wrong. He was an academic celebrity of the highest order, which brought both respect and freedom. His personal life, however, was a growing disaster: his (second) wife, Mimi, hated their Chicago life, which was an important part of the decision to move back to Cambridge, nearer to her family. But the move east didn't help much. Increasingly alienated at home, Black devoted more and more time to his work, branching off now in new directions. He began to work on generalizing CAPM to try to explain cycles in the economy: why, in a rational world, would there be periods of growth, followed by periods of contraction? This led him to a new theory of macroeconomics, which he called "general equilibrium." He also launched a crusade against the accounting industry, which he considered backward and unhelpful to investors.

But these other strands of his work were terribly received. It was as though Black had used up all his luck and timing with the options paper, and the string of other papers on derivatives and financial markets that followed it. His work on macroeconomics in particular was out of step with the times. Economists in the 1970s and 1980s were deeply engaged in an ongoing debate about economic regulation and monetary policy. On one side were the Chicagoans; on the other, the Keynesians, who favored government intervention throughout the economy. General equilibrium was a third way, thrust into a bipolar community. Black found himself attacked, and then ignored, by both sides. No one would publish his papers. His colleagues began to write him off as irrelevant. In less than a decade, he went from outsider to idol, and then back to outsider. By the early 1980s, Black was fed up with academia. He wanted out.

In December 1983, Robert Merton, Black's old collaborator from the Black-Scholes days, was doing consulting work for the investment bank Goldman Sachs. Merton was doing at Goldman what Black and Scholes had been doing at Wells Fargo back in 1970: bringing in the new ideas from academia and trying to implement them in a practical setting. In this capacity, he argued to Robert Rubin, then head of the Equities Division, that Goldman Sachs should hire a theorist, an academic of its own, at a high enough level of the company that the new ideas would have a chance to seep through the culture. Rubin was convinced, and so Merton went back to MIT, brainstorming who among their current crop of graduate students he would recommend for this important position. Merton asked Black for his advice and received a surprising answer: Black wanted the job himself. Three months later, Black left academia for a new job at Goldman Sachs, to organize a Quantitative Strategies Group in the Equities Division. Thus he became one of the first quants, a new kind of investment bank employee with an intensely quantitative and scientific focus, as interested in intellectual innovation as in making a big trade. Wall Street would never be the same.

On October 4, 1957, the Soviet Union launched *Sputnik*, the first man-made object to enter Earth's orbit. America panicked. Eisenhower immediately ordered the fledgling U.S. space program to schedule its own launch. The date was set for December 6. The event was televised live across the nation, as American scientists attempted to prove they were equal to the Soviets. Millions tuned in as the first American spaceship ignited on the launch pad, and then inched off the ground — for about four feet, before falling back to the tarmac and exploding. The performance was a humiliation for the American scientific establishment. Four years later, the Soviets did the Americans one better still, by propelling Yuri Gagarin into orbit and successfully launching the first manned spacecraft. Kennedy responded within the week by asking NASA to find a new challenge that the Americans could win. On May 25, 1961, Kennedy announced his commitment to put the first man on the moon.

Physics had been on the rise in the United States since World War II. But after *Sputnik* was launched, physics interest skyrocketed. About five hundred physics PhDs were awarded in 1958. By 1965, that number was closer to a thousand, and by 1969 it was over fifteen hundred. This rapid growth was in part a matter of nationalism: becoming a rocket scientist was a fine way to serve your country. But even more, it was a matter of funding. NASA's annual budget increased by a factor of seventy from 1958 to its peak in the mid-sixties. In 1966, NASA was given almost $6 billion — 4.5% of the total federal budget — to devote to basic science. Other government funding agencies, like the Department of Energy and the National Science Foundation, were also flush (though none could compete with NASA). Even mediocre graduates of mid-tier doctoral programs were guaranteed jobs in science, as either professors or government researchers. Physicists were in high demand.

On July 20, 1969, Neil Armstrong and Buzz Aldrin became the first men to set foot on the surface of the moon. America and its allies rejoiced — finally, an American victory in the space race. And almost immediately, the physics job market collapsed. As the space race accelerated, so too did America's commitment to the war in Vietnam. The success of the *Apollo 11* mission gave Nixon an excuse to divert funds from NASA and other research groups to the military effort. By 1971, NASA's budget was less than half of what it had been in 1966 (in real terms). Meanwhile, college enrollment began to drop, largely because the Baby Boom years were over. Once the "Boomers" had graduated, universities stopped hiring new faculty members.

Emanuel Derman was a South African physicist who experienced this funding roller coaster firsthand. He entered graduate school, at Columbia University, in 1966, at the high point of U.S. science funding. He worked on experimental particle physics—a field far from NASA's central interests, but a beneficiary of the uptick in government support for physics nonetheless. Like most graduate students, he slogged through, living on a small stipend and working long hours. The students he knew when he first arrived in graduate school went on to positions at universities around the country. But by the time Derman graduated, in 1973, there were no permanent jobs left. Derman, and other physicists who had done excellent work, were barely able to scrape together a series of temporary research positions. Derman spent two years at the University of Pennsylvania, followed by two years at Oxford, and then two years at Rockefeller University, in New York. By the end of the decade, he was ready to give up. He considered quitting physics for medical school but decided to go to Bell Labs and work as a programmer instead.

As the seventies droned on, the number of physics PhDs awarded in the United States dropped, to about one thousand per year. While this was significantly lower than the peak in 1968, it was still far more than the flailing job market could support. This meant that by the time Black moved to Goldman Sachs, in 1983, there were thousands of very talented men and women with graduate degrees in physics and related fields who were either unemployed or underemployed.

Black's move to Goldman Sachs coincided with another change, too. By 1983, options were a growing business, making people with training like Black's attractive on Wall Street. But bond trading—already a mainstay of the financial industry—was in the midst of a sea change. Beginning with the Carter administration in the late 1970s, the U.S. economy entered a period of high inflation and low growth that has subsequently been dubbed "stagflation." In response, Paul Volcker, the chairman of the Federal Reserve from 1979 through 1987, increased interest rates dramatically, so that the prime interest rate, the rate that determines how expensive it is for banks to lend to one another — and, by extension, to lend to consumers — reached an unprecedented level of 21.5%. Volcker was successful at reducing inflation, which he had under control by 1983. But this volatility in interest rates forever changed the previously sleepy bond industry. If banks couldn't borrow from one another for less than 20%, surely corporations and governments that were trying to issue bonds would need to pay even higher rates (since typically bonds are more risky than interbank loans). The so-called bond bores of the 1970s, traders who had chosen to work in the least exciting of the financial markets, now needed to cope with the most variable market of all. (Sherman McCoy, the star-crossed antihero of Tom Wolfe's novel *Bonfire of the Vanities*, was an eighties-era bond trader who took himself to be so important, given the changes in the bond markets during the late seventies and early eighties, that he privately called himself a "Master of the Universe." The name has stuck, now used to refer to Wall Street traders of all stripes.)

The success of the Black-Scholes model and other derivatives models during the 1970s inspired some economists to ask whether bonds could be modeled in a similar way to options. Soon, Black and others had realized that bonds themselves could be thought of as simple derivatives, with interest rates as the underlying asset. They began to develop modified versions of the Black-Scholes model to price bonds, based on the hypothesis that interest rates undergo a random walk.

Thus, Black arrived on Wall Street at a moment when derivatives, and derivative models, were proving increasingly important, in unexpected ways. Black's Quantitative Strategies Group at Goldman Sachs, as well as similar groups at other major banks, provided an answer to questions that many investment bankers, and especially bond traders, hadn't known how to ask. At the same time, there was a large pool of underemployed physicists who were ready to step in and follow Black's lead in changing financial practice. Once a few physicists and half-physicists had made their way to Wall Street, and once the usefulness of the ideas that Black had managed to translate from theory to practice was appreciated, the floodgates opened. Wall Street began hiring physicists by the hundreds.

Derman stayed at Bell Labs for five years. Starting in 1983, though, he began to get phone calls from headhunters sent from investment banks. He was unhappy enough at Bell Labs that he took these offers seriously, but when he finally received an offer from Goldman Sachs, he declined it on the advice of an acquaintance who had worked there before. But the world was changing. Derman found the next year at Bell Labs intolerable, and so when Wall Street came calling again, in 1985, he was ready to move. He decided to go with Goldman Sachs after all, and in December 1985 he made the leap. His job was in the Financial Services Group, which supported Goldman's bond traders. By the time he arrived, Black was already an institutional legend.

Both Thorp and Black based their options models on Osborne's random walk hypothesis, which amounted to assuming that rates of return are normally distributed. This might give you pause. After all, Mandelbrot argued throughout the 1960s that normal and log-normal distributions do not effectively account for extreme events, that markets are wildly random. Even if Mandelbrot's claim that rates of return are Lévy-stable distributed and thus do not have well-defined volatility is false — and most economists now believe it is — the weaker claim that market data exhibit fat tails still holds. Options models assign prices based on the probability that a stock will rise above (or drop below) a certain threshold — namely, the strike price for the option. If extreme market changes are more likely than Osborne's model predicts, neither Thorp's model nor the Black-Scholes model will get options prices right. In particular, they should undervalue options that would be exercised only if the market makes a dramatic move, so-called far-out-of-the-money options. A more realistic options model, meanwhile, should account for fat tails.

Mandelbrot left finance at the end of the 1960s, but he returned in the early 1990s. One of the reasons was that many financial practitioners were beginning to recognize the shortcomings of the Black-

Scholes model. Instrumental in this shift was the Black Monday stock market crash of 1987, during which world financial markets fell more than 20% literally overnight. Blame for the crash fell to a novel financial product based on options and the Black-Scholes model, known as portfolio insurance. Portfolio insurance was designed, and advertised, to curtail the risk of major losses. It was a kind of hedge that amounted to buying stocks and short selling stock market futures, the idea being that if stocks began to fall, the market futures would also fall, and so your short position would increase to offset your losses. The strategy was designed so that you wouldn't sell too many futures short, because that would eat into your profits if the market went up. Instead, you would program a computer to gradually sell your stocks if the market fell, and you would short just enough market futures to cover those losses.

When the market crashed in 1987, though, everyone with portfolio insurance tried to sell their stocks at the same time. The trouble with this was that there were no buyers — everyone was selling! This meant that the computers trying to execute the trades ended up selling at much lower prices than the people who had designed the portfolio insurance had expected, and the carefully calculated short positions in market futures did little to protect investors. (In fact, investors holding portfolio insurance tended to do better than those who didn't hold it; however, many people think the automated sell orders associated with portfolio insurance exacerbated the sell-off, and so everyone suffered because portfolio insurance was so prevalent.) The Black-Scholes-based calculations underlying portfolio insurance didn't anticipate the possibility of a crash, because the random walk model indicates that a major one-day drop like this wouldn't happen in a million years.

Several things happened in light of the crash. For one, many practitioners began to question the statistical predictions of the random walk model. This makes perfect sense — if your model says something is impossible, or virtually impossible, and then it happens, you need to start asking questions. But something else happened, too. Markets themselves seemed to change in the wake of the crash. Whereas in the years leading up to the crash the Black-Scholes model seemed to get options prices exactly right, in virtually all contexts and all markets, after the crash certain discrepancies began to appear. These discrepancies are often called the volatility smile because of their distinctive shape in certain graphs. The smile appeared suddenly and presented a major mystery for financial engineers in the early 1990s, when its prevalence was first recognized. Notably, Emanuel Derman came up with a way of modifying the Black-Scholes model to account for the volatility smile, though he never came up with a principled *reason* why the Black-Scholes model had stopped working.

Mandelbrot's work, however, offers a compelling explanation for the volatility smile. One way of interpreting the smile is as an indication that the market believes large shifts in prices are more likely than the Black-Scholes model assumes. This is just what Mandelbrot had been claiming all along: that probability distributions describing market returns have fat tails, which means that extreme events are more likely than one would predict based on a normal distribution. In other words, market forces seemed to have brought prices into line with Mandelbrot's theory. From the late 1980s on, Mandelbrot's work has been taken much more seriously by investment bankers.

There's an interesting, and rarely told, twist to the story of the rise and fall of Black-Scholes. The first major company to develop a quantitative strategy based on derivatives was a highly secretive Chicago firm called O'Connor and Associates. O'Connor was founded in 1977 by a pair of brothers named Ed and Bill O'Connor, who had made their fortune on grain futures, and Michael Greenbaum, a risk manager who had worked for them at First Options, an options clearinghouse the brothers ran. Greenbaum had majored in mathematics at Rensselaer Polytechnic Institute before joining First Options, and so he had some background with equations. He was one of the first people to realize that the new options exchange in Chicago offered a chance to make a killing, at least if you were mathematically sophisticated. He approached the O'Connor brothers with the idea of a new firm that would focus on options trading.

This much of the story is well known. But given the timing, many people assume that O'Connor was simply an early adopter of the Black-Scholes model. Not so. Greenbaum realized from the start that the assumptions underlying Black-Scholes weren't perfect, and that it was failing to properly account for extreme events. And so Greenbaum built a team of risk managers and mathematicians to figure out how to improve on the Black-Scholes model. One of O'Connor's first employees was an eighteen-year-old whiz kid named Clay Struve, who had worked for Greenbaum at First Options in a summer job, and who worked for Fischer Black as an undergraduate at MIT during the school year. During 1977 and 1978, Greenbaum, Struve, and a small team of proto-quants worked out a modified Black-Scholes model that took into account things like sudden jumps in prices, which can lead to fat tails.

O'Connor was famously successful, first in options and then in other derivatives — in part because the modified Black-Scholes model tended to outperform the standard one. Remarkably, according to Struve, O'Connor was aware of the volatility smile from very early on. That is, even before the crash of 1987, there were small, potentially exploitable discrepancies between the Black-Scholes model and market prices. Later, when the 1987 crash did occur, O'Connor survived.

There's another, deeper concern about the market revolution initiated by Black and his followers that many people worried about in 1987 and that has become quite stark in the wake of the most recent crisis. Take the 2008 crash as an example. During the financial meltdown, even sophisticated investors, such as the banks that produced securitized loans in the first place, appear to have been mistaken about how risky these products were. In other words, the models that were supposed to make these products risk-free for their manufacturers failed, utterly. Models have failed in other market disasters as well — perhaps most notably when Long-Term Capital Management (LTCM), a small private investment firm whose strategy team included Myron Scholes among others, imploded. LTCM had a successful run from its founding in 1994 until the early summer of 1998, when Russia defaulted on its state debts. Then, in just under four months, LTCM lost $4.6 billion. By September, its assets had disappeared. The firm was heavily invested in derivatives markets, with obligations to every major bank in the world, totaling about $1 trillion. Yet at the close of trading on September 22, its market positions were worth about $500 *million* — a tiny fraction of what they had been worth a few months before, and far too little to cover the company's loans. A feather's weight would have led to a default on hundreds of billions of dollars of debt, leading to an immediate international panic, had the government not stepped in to resolve the crisis.

The mathematical models underlying dynamic hedging strategies specifically, and derivatives trading more generally, are not perfect. Bachelier's, Osborne's, and Mandelbrot's stories go a long way toward making clear just why this is. Their models, and the models that have come since, are based on rigorous reasoning that, in a very real sense, cannot be wrong. But even the best mathematical models can be misapplied, often in subtle and difficult-to-detect ways. In order to make complicated financial markets tractable, Bachelier, Osborne, Thorp, Black, and even Mandelbrot introduced idealizations and often strong assumptions about how markets work. As Osborne in particular emphasized, the models that resulted were only as good as the assumptions that went in. Sometimes assumptions that are usually excellent quickly become lousy as market conditions change.

For this reason, the O'Connor story has an important moral. Many histories suggest that the 1987 crash rocked the financial world because it was so entirely unexpected — impossible to anticipate, in fact, given the prevailing market models. The sudden appearance of the volatility smile is taken as evidence that models can work for a while and then suddenly stop working, which in turn is supposed to undermine the reliability of the whole market-modeling enterprise. If models that work today can break tomorrow, with no warning and no explanation, why should anyone ever trust physicists on Wall Street? But this just isn't right. By carefully thinking through the simplest model and complicating it as appropriate — in essence, by accounting for fat tails — O'Connor was able to anticipate the conditions under which Black-Scholes would break down, and to adopt a strategy that allowed the firm to weather an event like the 1987 crash.

The story that I have told so far, from Bachelier to Black, goes a long way toward showing that financial modeling is an evolving process, one that proceeds in iterative fashion as mathematicians, statisticians, economists, and quite often physicists attempt to figure out the shortcomings of the best models and identify ways of improving them. In this, financial modeling is much like mathematical modeling in engineering and science more generally. Models fail. Sometimes we can anticipate when they will fail, as Greenbaum and Struve did; in other cases, we figure out what went wrong only as we are trying to put the pieces back together. This simple fact should urge caution as we develop and implement new modeling techniques, and as we continue to apply older ones. Still, if we have learned anything in the last three hundred years, it's that the basic methodological principles of scientific progress are the best ones we've got — and it would be foolish to abandon them just because they aren't always perfect.

What's more, since mathematical modeling in finance *is* an evolving process, we should fully expect that new methods can be developed that will begin to solve the problems that have plagued the models that have gotten us to where we are today. One part of this process has involved modifying the ideas that Black and Scholes introduced to financial practice to better accommodate Mandelbrot's observations about extreme events. But that's only the beginning. The final part of the book will show how models have continued to evolve outside of mainstream finance, as physicists have imported newer and more sophisticated ideas to finance and economics, identifying the problems with our current models and figuring out how to improve them. Black was instrumental in producing a new status quo on Wall Street, but his ideas were just the beginning of the era of financial innovation.

CHAPTER 6

The Prediction Company

When the Santa Fe Trail was first pioneered in 1822, it stretched from the westernmost edge of the United States — Independence, Missouri — through Comanche territory and into the then-Mexican state of Nuevo Mexico. From there it passed over the high plains of what is now eastern Colorado and then took the Glorieta Pass through the Sangre de Cristo Mountains, the southernmost subrange of the Rockies. To the southwest was the foot of the trail, the Palace of the Governors in the city of Santa Fe, the seat of Mexican power north of the Rio Grande. In front of the palace was the city's central market square, where traders from the United States displayed their goods. Twenty years after the first American trailblazers arrived in the city, the U.S. Army followed, battling through the Glorieta Pass and claiming the city and all of its surrounding territory as part of the newly annexed state of Texas.

A century and a half later, two men in their late thirties sat in a saloon at the end of the trail, long paved over and replaced by an interstate highway, sipping tequila. They were surrounded by younger men, chatting furiously. Outside, the park in the bustling market square was green from late-summer rains. Across the way, the Palace of the Governors sat as it always had, the oldest continuously used public building in North America. The square was surrounded by low-slung buildings, reddish brown and in the pueblo style, much as it was when the American army arrived in 1846. The men in the saloon were the newest traders to hang their sign in Santa Fe's historic market district. Down the road from the square, in a one-story adobe house on Griffin Street, a bank of state-of-the-art computers was humming, following the instructions set by the men before they left for their evening drink. The year was 1991. The men were in the prediction business.

The two graybeards — at least by the standards of the new field of nonlinear dynamics and chaos, which they had spent the last fifteen years helping to create — were James Doyne Farmer and Norman Packard. Until recently, Farmer had been head of the Complex Systems group at Los Alamos National Laboratory, the government lab most famous for having been the headquarters of the Manhattan Project. Packard, meanwhile, had just left a tenured position as associate professor of physics at the University of Illinois's flagship campus. Among the other men at the bar were former graduate students and recent PhDs, adventurers looking to follow Farmer and Packard as they blazed a new trail.

The new venture was a company, soon to be called the Prediction Company (though as they sat that evening on the Santa Fe market square, the company was still nameless). Their goal was to do the impossible: to predict the behavior of financial markets. If anyone could do it, it was this group. Between them, Farmer and Packard had three decades of experience in a subject known as nonlinear forecasting, an area of physics and applied mathematics (and increasingly other fields as well) that sought to identify predictive patterns in apparently random phenomena. In Packard's words, it involved identifying the order at "the edge of chaos," the small windows of time in which there was enough structure in a chaotic process to predict where a system would go next. The tools they used had been developed to predict things like how a turbulent fluid would behave in a narrow pipe. But Farmer and

Packard, and the half-dozen acolytes who had followed them to Santa Fe, believed they could predict far more than that.

As head of the Manhattan Project, J. Robert Oppenheimer was certainly the most important member of his family at Los Alamos. But he wasn't the only one. His kid brother, Frank, was also a physicist — and when the elder Oppenheimer took over work on the bomb, Frank pitched in, first at Lawrence Berkeley lab in California, and then at Oak Ridge in Tennessee, before finally joining his brother in New Mexico. Eight years younger than his famous brother, Frank arrived at Los Alamos just in time to help coordinate the Trinity test, the world's first nuclear detonation, which was staged in the middle of the Tularosa Basin in New Mexico on July 16, 1945. After the war, Robert appeared on the covers of *Time* and *Life*. He was the public spokesman for Cold War science in the United States, and for military restraint regarding the use of the nuclear technology he had helped develop. Frank was not quite so prominent, but even so, his military research landed him a job in the physics department at the University of Minnesota.

In 1947, J. Robert Oppenheimer was appointed director of both the Institute for Advanced Study in Princeton — possibly the most prestigious scientific research institute in the world — and the newly formed Atomic Energy Commission. The same year, the *Washington Times-Herald* reported that Frank Oppenheimer had been a member of the American Communist Party from 1937 to 1939. Eager as he was to continue in his brother's footsteps, 1947 was not a good year for a would-be nuclear physicist to be outed as a Communist. Frank initially denied the charges and appeared to have escaped with his reputation intact. But two years later, amid mass fear about Soviet nuclear research and the mishandling of the "atomic secret," Frank was called before the infamous House Un-American Activities Committee. Under oath and before Congress, he admitted that he and his wife had been members of the party for about three and a half years, pushed to political extremes during the Great Depression.

The confession was a newspaperman's dream. Frank Oppenheimer, brother of the American scientist-savior, was an admitted Communist. He was never convicted of a crime, nor was there any reason to think that he had compromised classified information. But during the heady and paranoid days of McCarthyism, the mere suggestion of Communist affiliation was enough to blacklist someone, no matter whose brother he was. Frank was forced to resign from his position at the University of Minnesota, and for more than a decade he was effectively strong-armed out of physics. Living on a substantial inheritance (sadly, he was forced to sell one of the van Goghs he'd inherited from his father), he and his wife bought a ranch in Colorado and made a new start as cattle farmers and homesteaders.

It was not until 1959 that McCarthyism had cooled enough that Frank Oppenheimer could get a job teaching physics at a research university, and even then it took the endorsements of a handful of Nobel and National Medal of Science laureates. Grateful to be back to work, he accepted a position at the University of Colorado. By now, though, the field had long outpaced him, so he limited himself to working on topics only indirectly connected to physics, such as science education.

It was at the University of Colorado that Oppenheimer met a young graduate student named Tom Ingerson. Ingerson had grown up in Texas and had gone on to major in physics at the University of California, Berkeley. He had come to Colorado to work on general relativity, the theory of gravitation that Einstein had proposed in 1915 as an alternative to Newton's theory. General relativity had brought fame and fortune to its discoverer, but it was overshadowed by the new quantum theory, which attracted far more attention and funds. This didn't seem to bother Ingerson, who was strong-willed and fiercely independent. He would work on what he liked.

In 1964, Ingerson began to think about finding a job in a physics department. In the 1960s, academia was an old boys' club in the strongest sense. Jobs at the top universities were filled by calling up famous physicists at famous schools and asking for recommendations — which were then given, in frank and certain terms. The "best men" from schools like Princeton, Harvard, and the University of Michigan were given the best jobs. Lesser men were dependent on the goodwill and reputation of their faculty, though personal connections and called-in favors were usually enough to find a job, especially during this, the heyday of the military-scientific-industrial era. Colorado may not have been in the very highest echelon, but it was up there, and a graduate could be reasonably assured of good employment. Unless, of course, he used the wrong person as a reference.

Ingerson didn't learn until many years later that his cardinal sin had been mentioning that Frank Oppenheimer would vouch for him. At the time, the physics community's uniform disinterest in his application seemed like a mystery to him. None of the employers he contacted wrote back to him until the very end of the school year, and then he heard from only a single school, the old New Mexico Territory's teaching college, newly retooled as Western New Mexico University. This was how a bright, independent-minded young physicist found himself in Silver City, New Mexico, the sole member of the local university's physics department.

Perched on the Continental Divide, Silver City was a paradigm Western mining town. Built in the wake of a major find by silver prospectors, it was in the middle of what was traditionally Apache territory. Trade and transport were difficult and dangerous, with regular attacks by regional tribes (and local bandits). In 1873, Billy the Kid, then just a teenager, settled in Silver City with his mother and brother — it was there that, in 1875, he was arrested for the first time, for stealing some cheese. Later that year, he would escape from a Silver City jail to begin his life as an outlaw, a fugitive from the Silver City sheriff. By the time Ingerson arrived, the days of cowboys and Indians were over. But Silver City was still a one-horse town. Resigned to make do with the cards he had inexplicably been dealt, Ingerson looked for ways to engage with the Silver City locals.

He started by volunteering with the local Boy Scout troop, which he thought might benefit from his experience as a teacher. It was at his first meeting, the same year that he moved to Silver City, that Ingerson met a pudgy twelve-year-old named Doyne Farmer. Silver City was filled with engineers, attracted by the mining industry. But a scientist was a rarity. Farmer didn't really know what a physicist did, yet he found Ingerson irresistible. Farmer decided at the meeting that whatever physics was, if Ingerson did it, then Farmer would do it, too.

He lingered afterward and then followed Ingerson home. On the way, Farmer announced his newfound career goal.

It was an unlikely friendship. But Farmer and Ingerson were kindred spirits, stuck for different reasons about as far away from the center of the scientific universe as they could be. For Ingerson, Farmer was a welcome diversion, a smart student ready to talk seriously about all sorts of scientific topics. For Farmer, though, Ingerson was pure inspiration. He changed his life.

Ingerson soon started a new group, which he called Explorer Post 114, with his home as clubhouse. The Explorer groups were a subsidiary of the Boy Scouts of America, intended for older children to learn by doing. Farmer was the inaugural member of Ingerson's group, but he was soon joined by others. The Explorers shared some features with the Boy Scouts—they went camping and hiking in the desert—but the real focus was on tinkering and building things, like ham radios and dirt bikes.

Officially, to join an Explorer post one needed to be at least fourteen years old. But one day in 1966, a younger boy was invited to come to a meeting. He had been asked to give a lecture on new radio technology, a topic on which he was apparently an expert. Though he was only twelve, the other explorers recognized Norman Packard as one of their own, and he was immediately welcomed into the group as the new electronics guru. Unlike Farmer, Packard had known he wanted to be a physicist from an early age. He seemed made for it. After all, it was his precocious expertise that earned him an invitation to the Explorers. Packard and Farmer quickly became friends.

Ingerson lasted for two more years in Silver City before he got a job at the University of Idaho. But in just four years, he had succeeded in shaping the lives of two men who would go on to become world-class physicists. When Ingerson left, Farmer was sixteen and a junior in high school (Packard was two years younger). Bored with Silver City and anxious to follow his friend to conquer new territory, Farmer decided to apply to the University of Idaho a year early. He got in, and instead of finishing high school in Silver City, he moved into Ingerson's attic in Moscow, Idaho, to start his career as a physicist. After a year in Idaho, though, Farmer was ready for bigger pastures. In 1970, he transferred to Stanford University. True to his ambitions, he majored in physics — laying the groundwork for a career that would change science, and finance, forever.

The ideas at the heart of Farmer's and Packard's work were first developed by a man named Edward Lorenz. As a young boy, Lorenz thought he wanted to be a mathematician. He had a clear talent for mathematics, and when it came time to select a major at Dartmouth, he had few doubts about what he would choose. He graduated in 1938 and went on to Harvard, planning to pursue a PhD. But World War II interfered with his plans: in 1942, he joined the U.S. Army Air Corps. His job was to predict the weather for Allied pilots. He was given this task because of his mathematical background, but at that point, at least, mathematics was of little use in weather forecasting, which was done more on the basis of gut feelings, rules of thumb, and brute luck. Lorenz was sure there was a better way — one that used sophisticated mathematics to make predictions. When he left the service in 1946, Lorenz decided to stick with meteorology. It was a place where he could put his training to productive use.

He went to MIT for a PhD in meteorology and stayed for the rest of his career — first as a graduate student, then as a staff meteorologist, and finally as a professor. He worked on many of the mainstream problems that meteorologists worked on, especially early in his career. But he had some unusual tastes. For one, based on his experience in the army, he maintained an interest in forecasting. This was considered quixotic at best by his colleagues; the poor state of forecasting technology had convinced many that forecasting technology was a fool's errand. Another oddity was that Lorenz thought computers — which, in the 1950s and 1960s, were little more than souped-up adding machines — could be useful in science, and especially in the study of complicated systems like the atmosphere. In particular, he thought that with a big enough computer and careful enough research, it would be possible to come up with a set of equations governing how things like storms and winds developed and changed. You could then use the computer to solve the equations in real time, keeping one step ahead of the actual weather to make accurate predictions long into the future.

Few of his colleagues were persuaded. As a first step, and as an attempt to show his fellow meteorologists that he wasn't crazy, Lorenz came up with a very simple model for wind. This model drew on the behavior of wind in the real world, but it was highly idealized, with twelve rules governing the way the wind would blow, and with no accounting for seasons, nightfall, or rain. Lorenz wrote a program using a primitive computer — a Royal McBee, one of the very first computers designed to be placed on a desk and operated by a single user — that would solve his model's equations and spit out a handful of numbers corresponding to the magnitude and direction of the prevailing winds as they changed over time. It wasn't a predictive model of the weather; it was more like a toy climate that incorporated atmospheric phenomena. But it was enough to convince at least some of his colleagues that this was something worth pursuing. Graduate students and junior faculty would come into his office daily to peer into Lorenz's imaginary world, taking bets on whether the wind would turn north or south, strengthen or weaken, on a given day.

At first, it seemed that Lorenz's model was a neat proof of concept. It even had some (limited) predictive power: certain patterns seemed to emerge over and over again, with enough regularity that a working meteorologist might be able to look for similar patterns in actual weather data. But the real discovery was an accident. One day, while reviewing his data, Lorenz decided he wanted to look at a stretch of weather more closely. He started the program, plugging in the values for the wind that corresponded to the beginning of the period he was interested in. If things were working as they should, the computer would run the calculations and come up with the same results he had seen before. He set the computer to work and went off for the afternoon.

When he returned a few hours later, it was obvious that something had gone wrong. The data on his screen looked nothing like the data he had seen the first time he had run the simulation with these same numbers. He checked the values he had entered — they were correct, exactly what had appeared on his printout. After poking around for a while, he concluded that the computer must be broken.

It was only later that he realized what had really happened. The computer contained enough memory to store six digits at a time. The state of Lorenz's mini world was summed up by a decimal with six figures, something like .452386. But he had set up the program to record only three digits, to save space on the printouts and make them more readable. So instead of .452386 (say), the computer printout would read .452. When he set up the computer to rerun the simulation, he had started with the shorter, rounded number instead of the full six-digit number that had fully described the state of the system during the first run-through.

This kind of rounding should not have mattered. Imagine you are trying to putt a golf ball. The hole you are aiming for is only slightly larger than the ball itself. And yet, if you miscalculate by a fraction of an inch, and you hit the ball a little too hard or a little too softly, or you aim a little to one side, you would still expect the ball to get close to the hole, even if it doesn't go in. If you are throwing a baseball, you would expect it to get pretty close to the catcher even if your arm doesn't extend exactly as you want it to, or even if your fingers slip slightly on the ball. This is how the physical world works: if two objects start in more or less the same physical state, they are going to do more or less the same thing and end up in very similar places. The world is an ordered place. Or at least that's what everyone thought before Lorenz accidentally discovered chaos.

Lorenz didn't call it chaos. That word came later, with the work of two physicists named James Yorke and Tien-Yien Li who wrote a paper called "Period Three Implies Chaos." Lorenz called his discovery "sensitive dependence on initial conditions," which, though much less sexy, is extremely descriptive, capturing the essence of chaotic behavior. Despite the fact that Lorenz's system was entirely deterministic, wholly governed by the laws of Lorenzian weather, extremely small differences in the state of the system at a given time would quickly explode into large differences later on. This observation, a result of one of the very first computer simulations in service of a scientific problem, contradicted every classical expectation regarding how things like weather worked. (Lorenz quickly showed that much simpler systems, such as pendulums and water wheels, things that you could build in your basement, also exhibited a sensitivity to initial conditions.)

The basic idea of chaos is summed up by another accidental contribution of Lorenz's: the so-called butterfly effect, which takes its name from a paper that Lorenz gave at the 1972 meeting of the American Association for the Advancement of Science called "Predictability: Does the Flap of a Butterfly's Wings in Brazil Set Off a Tornado in Texas?" (Lorenz never took credit for the title. He claimed one of the conference organizers came up with it when Lorenz forgot to submit one.)

Lorenz never answered the question asked in the title of his talk, but the implication was clear: a small change in initial conditions can have a huge impact on events down the road. But the real moral is that, even though chaotic systems are deterministic—in the sense that an infinitely precise description at any given instant can in principle lead to an accurate prediction—it is simply impossible to capture the state of the world with such precision. You can never account for all the flaps of all the butterflies across the globe. And even the tiniest errors will quickly explode into enormous differences. The result is that, even though weather is deterministic, it *seems* random because we can never know enough about butterflies.

Farmer finished his physics degree at Stanford in 1973, although not without a few bumps along the way (after his first year there, he had done poorly enough to be put on academic probation—after which he entertained the possibility of dropping out to open a smoothie shack in San Francisco or maybe smuggle motorcycles). By the end of his college years, however, Farmer had pulled himself together sufficiently to be admitted to a handful of graduate schools for astrophysics. A trip down the California coast was enough to make up Farmer's mind, however, and he decided to attend the new University of California campus in Santa Cruz. Packard, meanwhile, had gone to Reed College, in Portland, Oregon, a school famous for the independent spirit of its undergraduates.

During the summer of 1975, the year after Packard's junior year at Reed and Farmer's second year of graduate school, Packard and Farmer decided to try their hands at gambling. They had explored the idea independently, Farmer through reading A. H. Morehead's *Complete Guide to Winning Poker,* and Packard by reading Ed Thorp's *Beat the Dealer.* With their analytic minds and disdain for authority, both men found gambling systems had a certain appeal. They could make money without doing work—and at least in the blackjack case, they could do it by being smarter than everyone else. It was a romantic idea. The trouble was in the execution.

Packard studied Thorp's system carefully and then, along with a friend from Reed named Jack Biles, he took it to Vegas. They kept careful track of their winnings and losses—and observed an awful lot of wins. Day after day, they would record profits. They would switch to higher-stakes tables as their accumulated capital increased, and the profits would soar even higher. But then something happened. No matter how much success they had, there would always be a losing streak that would bring them back to zero. In the end, they barely broke even. It was only at the very end of a summer of gambling that they realized they were being cheated. In the years since Thorp's card-counting system had first been introduced, casinos had become very good at identifying—and foiling—card counters, often by simple methods like crooked dealing.

Farmer, meanwhile, had memorized Morehead's book. But he had never played poker before reading it, so even though he knew what to do in any given situation, he didn't know how to shuffle cards or handle chips. He dealt like a kindergartener. But the poor mechanics ultimately worked to his advantage: he looked like an easy mark. Playing in the card rooms of Missoula, Montana, under the alias "New Mexico Clem," Farmer and a friend from Idaho—an accomplice from the motorcycle-smuggling scheme named Dan Browne—cleaned up against the Missoula cowboys. Browne, a more seasoned player who had paid his way through college by gambling in Spokane, Washington, marveled at Farmer's unlikely success.

At the end of the summer, Farmer and Packard decided to meet up to compare notes on their gambling adventures. Farmer had good news to report: you could make a killing in poker, if you just played by the book. Packard's experience was less auspicious. But in place of blackjack winnings, he brought something even better: a new idea for a gambling system. Prompted in part by some cryptic remarks that Thorp had made at the end of his book, Packard convinced himself that another game could be beaten more effectively than blackjack (and with less likelihood for casino shenanigans). Packard, like Thorp before him, had an idea about roulette.

Farmer was skeptical, but Packard was persistent and finally convinced Farmer to think about it. Soon enough, Farmer was on board. He, Packard, and Biles spent three days thinking about the problem, working out some initial calculations and getting excited about their newest project. By the time Farmer had to go back to Santa Cruz, the three men had decided to pursue the project. They would build a computer to beat roulette.

In the fall of 1975, Farmer was starting his third year of graduate school. He was supposed to be settling on a dissertation topic and beginning research in astrophysics. Instead, he and Browne began running experiments on a roulette wheel they had bought in Reno, at Paul's Gaming Devices, the manufacturer rumored to provide the regulation wheels used in Reno and Las Vegas. (Farmer's thesis advisor, a man named George Blumenthal, had enjoyed his own run as a would-be card counter in Las Vegas. He was tickled enough by Farmer's project to look the other way as Farmer's academic research stalled — in fact, after reviewing Farmer's calculations, he even suggested that there might be a physics dissertation lurking in the roulette project.) Packard and Biles, meanwhile, were back in Portland, working on an electronic clock that could take precise measurements of the ball traveling around the wheel. Along with his work on the roulette project, Packard was finishing college and applying to graduate school. Santa Cruz was at the top of his list. At this stage, even though Packard knew Thorp had thought about beating roulette, no one in the group knew anything about Thorp's calculations, or about the computer that Thorp and Shannon had tested in Las Vegas. They were reinventing the wheel.

At the end of that academic year, in the spring of 1976, the four gambling men met up in Santa Cruz to put their work together and make a plan for the summer. One of their first pieces of business was to settle on a name for the group. Farmer had recently stumbled on a new word, *eudaemonia,* while flipping through the dictionary. Central to the ethics of the ancient Greek philosopher Aristotle, eudaemonia was a state of ideal human flourishing. The roulette group took the name Eudaemonic Enterprises, and the members referred to themselves as Eudaemons (Greek for "good spirits"). They rented a professor's house for the summer and built a tinkerer's lab, assembling electronics and running experiments on roulette wheels. The Eudaemons independently arrived at the same basic strategy that Thorp and Shannon had used, with two people working the game, one timing the wheel and the other making bets. Ingerson's legacy was manifest in Farmer and Packard's conviction that they could build anything. The Eudaemons were an only slightly more grown-up version of the Explorer Post 114 (indeed, Ingerson later helped the group in Vegas when they tried to put the scheme into action).

The original four were soon joined by another physicist named John Boyd and a friend of Farmer's from his undergraduate days, Steve Lawton. Lawton was a humanist, a specialist in utopian literature. His role was to organize a reading group on political fiction. From the start, the group was devoted to a revolutionary mindset. Over the years, as they continued to work on roulette, more and more people joined — gamblers, physicists, computer programmers, utopians. The group thought of themselves as Yippies, members of the countercultural movement founded by Abbie Hoffman and others in 1967 and devoted to undermining the status quo through anarchic pranks they called "Groucho Marxism." For the Eudaemons, the roulette project was a way to beat the Man and take his money — money they planned to use to build a commune on the Washington coast.

Thorp and Shannon never had much luck with their roulette adventure, on account of frayed wires and nerves. The Eudaemons did better, plugging away at the problem for the better part of five years. Not that they didn't have their own share of hardware problems. Instead of an earpiece like Thorp wore, the Eudaemons' first generation of technology sent signals via a vibrating magnet attached to the bettor's torso, hidden by clothes. One night, the wires on Farmer's magnet kept coming undone, burning his skin whenever the signal arrived. Every ten minutes he had to jump up from the table and announce something like "Boy, have I got the runs today!" on his way to the men's room to fix the equipment (this continued until the pit boss followed him in and sat in the next stall until Farmer decided to call it quits for the night). But by the summer of 1978 the computers were running well enough that the team took them to Vegas — and started to profit.

Meanwhile, as the team at Eudaemonic Enterprises continued work on building a better bettor, Farmer, Packard, and some of the others in the group began thinking more about the physics at the heart of the project. They had derived the equations they needed to predict roulette. But thinking about roulette had piqued their interest in a more general problem. Roulette is an example of a dynamical system that exhibits some pretty funky behavior. Most importantly, where the ball lands is sensitive to the initial conditions — much like the weather system Lorenz discovered. Working out how to use computers to solve the differential equations necessary to predict roulette had unwittingly put Farmer and Packard at the cutting edge of the newest research in chaos theory. Farmer's advisor was right that there was a dissertation in the roulette calculations. What he didn't know was that the dissertation would be part of a rising tide of ideas that would usher in a new age of physics.

In 1977, some of the physicists working on Eudaemonic Enterprises (Farmer and Packard, along with an undergraduate named James Crutchfield and an older graduate student named Robert Shaw) started an informal research group called by turns the Dynamical Systems Collective and the Chaos Cabal. Shaw threw out a nearly finished dissertation to start working on chaos theory full-time; Farmer officially switched away from astrophysics. By the late 1970s, a great deal had been done on chaos theory. Lorenz had discovered many of the basic principles and had then come up with simple examples of chaotic systems and described how they behaved. He was the first person to recognize that there is a kind of order in chaotic systems: if you draw pictures of the paths traced by objects obeying differential equations, they tend to settle down into regular patterns. These patterns are called attractors, because they tend to attract the paths of the objects. In roulette, for instance, the attractors correspond to the pockets of the wheel: whatever trajectory the ball takes, in the long run it will settle down into one of these states. But for other systems, the attractors can be much more complicated. A major contribution to the study of chaos theory was the realization that if a system is chaotic, these attractors have a highly intricate fractal structure.

But despite these foundations, the subject was still young. Work had been done in fits and starts, without any real research center. Normally, graduate work in physics is a collaboration among graduate students, young postdoctoral researchers, and a professor. But chaos theory was still so new that these kinds of research groups didn't yet exist. You couldn't go to graduate school to study chaos theory. The Dynamical Systems Collective was an attempt to fix this, by pulling its members through graduate school by their bootstraps. Some of the faculty at Santa Cruz were skeptical about this divergence from the traditional academic curriculum. But the department was new and open to novel ideas, and enough professors were supportive that the four initial members were permitted to guide themselves, collectively, to PhDs in chaos theory.

From the very start, prompted perhaps by the roulette experience, the Dynamical Systems Collective was interested in prediction. It was a novel way of thinking about chaotic systems, which most people were interested in precisely because they seemed so *unpredictable.* The collective's most important paper, published in 1980, showed how you could use a stream of data from, say, a sensor placed in the middle of a pipe with water flowing through it to reconstruct what the attractor for the system would have to be. And once you had the attractor, an essential part of trying to understand how a chaotic system would behave over time, you could begin to make some predictions. Previously, attractors were understood as a theoretical tool, something you could get only by solving equations. Packard, Farmer, Shaw, and Crutchfield showed that, in fact, you could figure out this important feature empirically, by looking at how the system actually behaved.

The Dynamical Systems Collective lasted for four years, during which time it made seminal advances in chaos theory and managed to turn years of thinking about roulette into respectable science. But the Eudaemons couldn't stay in graduate school forever. Farmer graduated in 1981 and immediately went to Los Alamos. Packard left the following year, to take a postdoctoral position in France. Both men were on the verge of turning thirty when they left school. Eudaemonic Enterprises was making money from roulette, but it was ultimately a state of mind, not a way to earn a living.

It was a miracle that either Farmer or Packard got academic jobs, with degrees in chaos in the early 1980s, when few physicists knew what the new theory of dynamical systems was all about, and even fewer recognized it as something worth pursuing. Los Alamos, like Santa Cruz, was far ahead of its time, and Farmer was fortunate to find himself at the center of research in the new field. (Packard had similar luck. After his postdoctoral year in France, he landed positions at the Institute for Advanced Study, in Princeton, New Jersey, and the Center for Complex Systems Research, at the University of Illinois, the other two hotbeds of complex systems research.) Things got even better for Farmer in 1984, when a group of senior scientists at the lab launched a new research center devoted to the study of complex systems, including chaos. The center was called the Santa Fe Institute. Physics would play a central part in the Santa Fe Institute's research, but the center was designed to be essentially interdisciplinary. Complex systems and chaos arose in physics, in meteorology, in biology, in computer science — and also, the Santa Fe researchers soon realized, in economics.

One theme that characterized much of the research in complexity and chaos during the early 1980s was the idea that simple large-scale structures can emerge from underlying processes that don't seem to have that structure. To take an example from atmospheric physics, consider that the atmosphere, at the smallest scale, consists of a bunch of gas particles bumping around in the sky. And yet, when one steps back, these mindless particles somehow organize themselves into hurricanes. Similar phenomena occur in biology. Individual ants seem to behave in pretty simple ways, foraging for food, following pheromone trails, building nests. And yet, when one takes these simple actions and interactions in aggregate, they form a colony, something that appears to be more than the sum of its parts. As a whole, an ant colony even appears to be able to adapt to changes in its environment, or the deaths of individual ants. Once these ideas were in the air at Santa Fe, it was a natural leap to ask if the economies of nations and the behavior of markets could also be understood as the collective action of individual people.

The Santa Fe Institute hosted its first conference on economics, entitled "International Finance as a Complex System," in 1986. Farmer, who at this point was the head of the Complex Systems research group at Los Alamos, was one of a small handful of scientists who were asked to speak. It was his first exposure to economics. The other speakers were from various banks and business schools. These bankers stood up and explained their models to a group of stunned scientists who found the financial models almost childishly simple. The bankers, meanwhile, walked away thinking that they had heard the siren call of the future, though they had virtually no understanding of what was being said. Excited, they urged the institute to host a follow-up conference and invite various luminaries from economics departments at top universities.

The idea behind the second conference was that even if the financiers couldn't follow the latest advances in physics and computer science, surely the professional economists would be able to. Unfortunately, things didn't go as planned. Farmer and Packard both spoke, as did various other Santa Fe Institute researchers. The economists, likewise, made their presentations. But there wasn't much communication. The two groups were coming from two radically different cultures and taking too many different things for granted. The physicists thought the economists were making everything much too simple. The economists thought the physicists were talking nonsense. The great synthesis of disciplines never occurred.

Undeterred, the institute tried a third time in February 1991. This time, though, the economists stayed at home. Instead, the institute invited practitioners from the banks and investment houses that actually ran the world's financial markets. The tone of the conference was much more practical and focused on how to create models, test them, and use them to develop trading strategies. The traders proved much less defensive than the economists, and by the end of the conference each group had gained an appreciation of what the other had to offer. Farmer and Packard, in particular, left with a clearer sense of how practical trading strategies worked. They also left with the conviction that they could do better. A month later, they gave notice to their respective employers. It was time to enter the fray.

Building a company is different from building a radio or a motorcycle engine, or even a computer to beat roulette. But many of the same skills prove useful: the vision to see how to pull the pieces together in a new way; a tolerance for tinkering with something until you can make it work; unflagging persistence. Making something new is addictive, which might be why so many entrepreneurs are engineers and scientists.

Farmer and Packard were also motivated by a strong antiestablishmentarian bent stretching back to their days as Eudaemons. The new company wasn't designed as a first step into the financial world — it was part of a plan to upend it, to take Wall Street for all it was worth by being a little smarter, a little more conniving, than the suits. It was a company founded in much the same spirit as the roulette project, a Yippie adventure and a return to a culture of pure research and no rules. Farmer wore an EAT THE RICH T-shirt to the new company's first formal meeting, in March 1991.

But there was more at stake here than in roulette. Farmer and Packard wanted the project to work, and they were willing to consider the possibility that real business acumen could be useful to them. So they brought in Jim McGill, a former physicist turned entrepreneur, as a third partner. In 1978, McGill had founded a company called Digital Sound Corp., which specialized in the kinds of microchips necessary to process data from electric musical instruments and microphones, and then later branched out into voice-mail devices. McGill was, at least nominally, the CEO of the Prediction Company, the business face of their Birkenstock-and-blue-jean outfit. Farmer and Packard were perfectly adept at imagining what they would do with, say, a hundred million in capital. McGill's job was to find someone to give it to them.

McGill would be the difference between the Prediction Company and a rerun of Eudaemonic Enterprises.

It quickly turned out, however, that finding would-be investors wasn't as difficult as the founders imagined it would be. Farmer and Packard had earned reputations during the days of the Santa Fe Institute's economics conferences. When rumors began flying that Farmer and Packard were leaving academia to take on Wall Street, some influential people took notice. Farmer had to buy a new suit to look presentable for meetings at places like Bank of America and Salomon Brothers. Things got even better after the *New York Times Magazine* ran a cover article called "Defining the New Plowshares Those Old Swords Will Make," on how physicists, who had largely been absorbed into the military-industrial complex in the wake of World War II, were branching out as the Cold War came to an end. The article led with the Prediction Company—a perfect tie-in, given Farmer's history with Los Alamos. After it appeared in print, hundreds of suitors began to call, from rich oil men to Wall Street banks.

The trouble wasn't getting money. It was what the would-be investors wanted in exchange. Some of the Wall Street outfits were thrilled with the idea of starting a hedge fund based on the Prediction Company's ideas. But Farmer and Packard didn't like the idea of traveling the country trying to raise capital, which they would need to do if they were managing a hedge fund. Ideally, they wanted seed money so they could focus on developing the science. Other companies wanted to buy the Prediction Company outright—equally unappealing to a group of men who had just made up their minds to break out of the rat race and start their own business. Some companies were willing to put up capital in exchange for a portion of the proceeds, but they wanted more than just a return on their investment. For instance, David Shaw, a former computer science professor at Columbia who had started his own hedge fund, D. E. Shaw & Co., in 1988, wanted to own the company's intellectual property in exchange for a few years' worth of seed money.

Many of these offers were appealing. But Farmer and Packard continued to balk. Nothing felt right. Unfortunately, they couldn't run an investment firm on the backs of their personal checking accounts forever. As the company's one-year anniversary approached, in March 1992, the pressure was on to find a deal.

It is tempting to say that Farmer, Packard, and their Prediction Company collaborators "used chaos theory to predict the markets" or something along those lines. In fact, this is how their enterprise is usually characterized. But that isn't quite right. Farmer and Packard didn't use chaos theory as a meteorologist or a physicist might. They didn't do things such as attempt to find the fractal geometry underlying markets, or derive the deterministic laws that govern financial systems.

Instead, the fifteen years that Farmer and Packard spent working on chaos theory gave them an unprecedented (by 1991 standards) understanding of how complex systems work, and the ability to use computers and mathematics in ways that someone trained in economics (or even in most areas of physics) would never have imagined possible. Their experience with chaos theory helped them appreciate how regular patterns — patterns with real predictive power — could be masked by the appearance of randomness. Their experience also showed them how to apply the right statistical measures to identify truly predictive patterns, how to test data against their models of market behavior, and finally how to figure out when those models were no longer doing their job. They were at ease with the statistical properties of fat-tailed distributions and wild randomness, which are characteristic of complex systems in physics as well as financial markets. This meant that they could easily apply some of Mandelbrot's ideas for risk management in ways that people with more traditional economics training could not.

As far as the Prediction Company was concerned, markets might be chaotic, or not. There might be various degrees of randomness in market behavior. Markets might be governed by simple laws, or by enormously complicated laws, or by laws that changed so fast that they might as well not have been there at all. What the Predictors were doing, rather, was trying to extract small amounts of information from a great deal of noise. It was a search for regularities of the same sort that lots of investors look for: how markets react to economic news like interest rates or employment numbers, how changes in one market manifest themselves in others, how the performances of different industries are intertwined.

One strategy they used was something called statistical arbitrage, which works by betting that certain statistical properties of stocks will tend to return even if they disappear briefly. The classic example is pairs trading. Pairs trading works by observing that some companies' stock prices are usually closely correlated. Consider Pepsi and Coca-Cola. Virtually any news that isn't company-specific is likely to affect Pepsi's products in just the same way as Coca-Cola's, which means that the two stock prices usually track one another. But changes in the two companies' prices don't always occur simultaneously, so sometimes the prices get out of whack compared to their long-term behavior. If Pepsi goes up a little bit but Coca-Cola doesn't, upsetting the usual relationship, you buy Coca-Cola and sell Pepsi because you have good reason to think that the two prices will soon revert to normal. Farmer and Packard didn't come up with pairs trading — it was largely pioneered in the 1980s at Morgan Stanley, by an astrophysicist named Nunzio Tartaglia and a computer scientist named Gerry Bamberger — but they did bring a new level of rigor and sophistication to the identification and testing of the statistical relationships underlying the strategy.

This sophistication was purely a function of the tools that Farmer and Packard were able to import from their days in physics. For instance, as a physicist, Packard was at the very forefront of research in a variety of computer programs known as genetic algorithms. (An algorithm is just a set of instructions that can be used to solve a particular problem.) Suppose you are trying to identify the ideal conditions under which to perform some experiment. A traditional approach might involve a long search for the perfect answer. This could take many forms, but it would be a direct attack. Genetic algorithms, on the other hand, approach such problems indirectly. You start with a whole bunch of would-be solutions, a wide variety of possible experimental configurations, say, which then compete with one another, like animals vying for resources. The most successful possible solutions are then broken up and recombined in novel ways to produce a second generation of solutions, which are allowed to compete again. And so on. It's survival of the fittest, where fitness is determined by some standard of optimality, such as how well an experiment would work under a given set of conditions. It turns out that in many cases, genetic algorithms find optimal or nearly optimal solutions to difficult physics problems very quickly.

Physicists in general, and Farmer and Packard especially, have developed many kinds of optimization algorithms that, by different means, accomplish the same goals as genetic algorithms, with different algorithms carefully tailored to different tasks. These algorithms are pattern sleuths: they comb through data, testing millions of models at a time, searching for predictive signals.

But there's nothing special about physics problems, as far as these algorithms are concerned. They can be applied to any number of different areas — including finance. Suppose you have discovered some strange statistical behavior relating the currency market for Japanese yen with the market for rice futures. It might seem sufficient to observe that if yen go up, then so do rice futures prices. You would then buy rice futures whenever you noticed yen ticking upward. Or else, suppose you have an idea for a possible pairs trade, such as with Pepsi and Coca-Cola.

Notice that in these cases, the basic strategy is clear. But there are all sorts of possibilities compatible with that basic strategy. To be perfectly scientific about the problem, you would want to figure out just how closely correlated the two prices are, and whether the degree of correlation varies with other market conditions. You would also want to think about how much rice to buy and how to time your purchase to be maximally certain that yen were really going up. But trying to come up with a way of relating all of these variables in an optimal way from scratch would be an enormously time-consuming and difficult process, and you could never be sure you'd gotten it right. In the meantime, your opportunity would pass. But if you used a genetic algorithm, you could let thousands of closely related models and trading strategies based on the supposed connection between yen and rice compete with one another. You would soon arrive at an optimal, or nearly optimal, strategy. This is a variety of forecasting, but it doesn't require you to come up with some complete chaos-theoretic description of markets. It's much more piecemeal than that.

Another one of the Prediction Company's ideas was to use many different models at once, each based on different simplified assumptions about the statistical properties of different assets. Farmer and Packard developed algorithms that allowed the different models to "vote" on trades — and then they adopted a strategy only if their models were able to form a consensus that it would likely be successful. Voting may not sound as if it has anything to do with physics, but the underlying idea comes right from Farmer's and Packard's days studying complex systems. Allowing many different models to vote identifies which trading strategies are robust, in the sense that they aren't sensitive to the special details of a particular model. There is a close connection between searching for robust strategies and searching for attractors in a complex system, since attractors are independent of initial conditions.

This kind of modeling, where one uses algorithmic methods to identify optimal strategies, is often called "black box" modeling in the financial industry. Black box models are very different from models like Black-Scholes and its predecessors, whose inner workings are not only transparent but often provide deep insights into why the models (should) work. Black box models are much more opaque, and as a result they are often scarier, especially to people who don't understand where they come from or why they should be trusted. Black box models were occasionally used before the Prediction Company came along, but the Prediction Company was one of the very first companies to build an entire business model based on them. It was a whole new way of thinking about trading.

Almost a year into the new company, the senior partners weren't making any money. An investment firm needs something to invest. Farmer, Packard, and McGill could go only so long without bringing home paychecks — and to make matters worse, they had been funding their team of graduate students and computer hackers out of their own pockets for eight months, since everyone had taken up residence at the

Griffin Street office in July 1991. The time for being choosy was coming to an end. The partners knew they didn't want to sell the company so soon into the adventure, but the idea of being someone else's hedge fund was starting to look appealing. At least they'd have capital, and they would be (more or less) independent. They had spent months interviewing possible partners, and at this point it was hard to imagine a better solution.

And then, in early March 1992, a miracle happened. Farmer had been invited to give a presentation at an annual computer conference. He had reluctantly agreed to attend, on the basis that Silicon Valley investors would be there and they might be willing to offer some no-strings-attached financing. He gave a talk on the role of computers in prediction, which generated a lot of questions. Afterward, as he was packing up his slides, a man in a suit approached him. He introduced himself as Craig Heimark, a partner at O'Connor and Associates — the firm that had made its first fortune by successfully modifying the Black-Scholes equation to account for fat-tailed distributions, under the guidance of Michael Greenbaum and Clay Struve. By 1991, it was one of the biggest players in the Chicago commodities markets, with a focus on high-tech derivatives trading. The company had six hundred employees and billions of dollars under management. O'Connor wasn't using nonlinear forecasting, and the Prediction Company wasn't interested in derivatives. But nonetheless, O'Connor and Associates were the Predictors' kind of people. In fact, one of O'Connor's recent hires had been a friend and fellow researcher back in Farmer's and Packard's academic days.

Shortly after Farmer and Heimark met, Farmer received a phone call from another O'Connor partner, named David Weinberger. Weinberger had been one of the very first quants, leaving a teaching job in operations research (essentially, a branch of applied mathematics) at Yale to work for Goldman Sachs in 1976, even before Black arrived. He'd moved to O'Connor in 1983, to help that company come up with new strategies as more and more companies got on the Black-Scholes bandwagon. He was one of the few people in the industry, even in 1991, who both was high powered enough to make a deal and also spoke the language of the scientists running the Prediction Company. He called on a Friday afternoon, from Chicago. On Saturday morning, he was sitting in the Griffin Street office.

O'Connor turned out to be just the kind of firm that the Prediction Company wanted to work with — in large part because the people working at O'Connor were able to understand what Farmer and Packard were doing well enough to evaluate it themselves. Under the deal they ultimately negotiated, the Prediction Company maintained its independence. O'Connor put up the investment capital, in exchange for the majority of the proceeds; it also fronted the Prediction Company the funds it so desperately needed in order to pay salaries and buy equipment in the meantime.

The deal with O'Connor seemed perfect at the time. But it turned out to be even better than the Prediction Company founders had hoped. When O'Connor came knocking on the Prediction Company's door, it already had a long-running partnership with Swiss Bank Corporation (SBC), a nearly century-and-a-half-old Swiss bank. And then, in 1992, before the ink was dry on O'Connor's deal with the Prediction Company, SBC announced its intention to buy O'Connor outright. The Prediction Company found itself in a partnership negotiated with its kindred spirits at O'Connor but funded by the much deeper pockets of SBC. Weinberger was given a top management position at SBC and continued as the principal liaison for the Prediction Company. It was an ideal arrangement. The Predictors had hit the big time.

In 1998, SBC merged with the still-larger Union Bank of Switzerland to form UBS, one of the largest banks in the world. Despite the size difference, however, most of the senior positions at UBS went to former SBC managers and the relationship with the Prediction Company was maintained.

The Prediction Company, following the O'Connor tradition as a secretive high-tech firm, never released any metrics of its success publicly — and none of the former principals or board members with whom I spoke were authorized to share any concrete information. This might seem suspicious. After all, if you're successful, why hide it? Here, though, the opposite is the case: on Wall Street, success breeds imitation, and the more firms there are implementing a strategy, the less profitable it is for anyone. There are some indications, however, that the Prediction Company has been wildly successful. As one former board member I spoke with pointed out, it is still an active subsidiary of UBS, after more than a decade. Another knowledgeable source told me that, over the firm's first fifteen years, its risk-adjusted return was almost one hundred times larger than the S&P 500 return over the same period.

Farmer stayed with the firm for about a decade before his passion for research lured him back to academia. He took a position at the Santa Fe Institute as a full-time researcher in 1999. Packard stayed with the company for a few more years, serving as CEO until 2003, when he left to start a new company, called ProtoLife. By the time they left, they had made their point: a firm grasp of statistics and a little creative reappropriation of tools from physics were enough to beat the Man. It was time to tackle a new set of problems.

Black box models, and more generally "algorithmic trading," have taken much of the backlash against quantitative finance in the period since the 2007–2008 financial crisis. The negative press is not undeserved. Black box models often work, but by definition it is impossible to pinpoint *why* they work, or to fully predict when they are going to fail. This means that black box modelers don't have the luxury of being able to guess when the assumptions that have gone into their models are going to turn bad. In place of this sort of theoretical backing, the reliability of black box models has to be constantly tested by statistical methods, to determine the extent to which they continue to do what they are intended to do. This can make them seem risky, and in some cases, if used injudiciously, they really *are* risky. They are easy to abuse, since one can convince oneself that a model that has worked before is a kind of magical device that will continue to work, come what may.

In the end, though, data outclass theory. This means that no matter how good the theoretical backing for your (non–black box) model, you ultimately need to evaluate it on the basis of how well it performs. Even the most transparent models need to be constantly tested by just the same kinds of statistical methods that are used to evaluate black box models. The clearest example of why this is so can be found by looking at the failure of the Black-Scholes model to account for the volatility smile in the aftermath of the 1987 crash. Theoretical backing for a model can be a double-edged sword: on the one hand, it can help guide practitioners who are trying to understand the limits of the model; conversely, it can lull you into a sense of false confidence that, because you have some theoretical justification for a model, the model must be right. Unfortunately science doesn't work this way. And from this latter point of view, black box models have an advantage over other, more theoretically transparent models, because one is effectively forced to evaluate their effectiveness on the basis of their actual success, not on one's beliefs about what *ought* to be successful.

There's another worry about black box models, above and beyond their opaqueness. All of the physicists whose work I have discussed thus far, from Bachelier to Black, have argued that markets are unpredictable. Purely random. The only disputes concern the nature of the randomness, and whether they are well enough behaved to be treated by normal distributions. In the years since it was first observed by Bachelier and Osborne, the idea that markets are unpredictable has been elevated to a central tenet of mainstream financial theory, under the umbrella of the efficient market hypothesis.

And yet, the Prediction Company, and dozens of other black box trading groups that have sprung up subsequently, purports to predict how the market will behave, over short periods of time and under special circumstances. The Prediction Company, at least, never worked with derivatives — its models attempted to predict how markets would behave directly, in just the way that many economists (and plenty of investors) would have supposed was impossible. Nonetheless, it was successful.

It's reasonable to be skeptical about the company's success. Investing can often come down to luck. That markets are random is not just conventional wisdom in economics departments. There's an enormous amount of statistical evidence to support it. Then again, perhaps the idea that markets are random because they are efficient — in the sense that market prices quickly change to account for all available information concerning the expected future performance of a stock — is not necessarily in conflict with the Prediction Company's success. It sounds like a paradox. But think about the basis for the efficient market hypothesis. The standard argument goes something like this: Suppose that there was some way to game the markets; that is, suppose that there was some reliable way to predict how prices are going to change over time. Then investors would quickly try to capitalize on that information. If markets are always at a local high in the last week of May, or if they always drop on the Monday following a Giants victory, then as soon as the pattern gets noticed, sophisticated investors will start selling stocks at the end of May and buying them as soon as the Giants win — with the result that prices will drop at the end of May and rise on Mondays after Giants victories, essentially washing out the pattern. Sure enough, every time an economist appears to find an anomalous pattern in market behavior, it seems to correct itself before the next study can be done to confirm it.

Fair enough. This kind of reasoning might make you think that even if markets somehow got out of whack, there are internal processes that would quickly push them back into shape. (Of course, one of the major reasons to think that the efficient markets hypothesis is deeply flawed is the apparent presence of speculative bubbles and market crashes. Whether these kinds of large-scale anomalies, where prices seem to become unmoored, are predictable is the subject of the next chapter. Here I am thinking of smaller-scale deviations from perfect efficiency, supposing that such a thing exists.) But what are these internal processes? Well, they involve the actions of so-called sophisticated investors, people who are quick to identify certain patterns and then adopt trading strategies designed to exploit those patterns. These sophisticated investors are what *make* the markets random, at least according to the standard line. But they do so by correctly identifying predictive patterns when they arise. Such patterns might disappear quickly. But if you're the *first* person to notice such a pattern, the argument about self-correcting markets doesn't apply.

What does this mean? It means that even if you take the standard line on efficient markets seriously, there is still a place for sophisticated investors to profit. You just need to be the *most* sophisticated investor, the one most carefully attuned to market patterns, and the one best equipped to find ways to turn patterns into profit. And for *this* task, a few decades of experience in extracting information from chaotic systems plus a room full of supercomputers could be a big help. In other words, the Prediction Company succeeded by figuring out how to be the most sophisticated investor as often as possible.

Of course, not everyone buys the idea that markets are efficient. Farmer, for one, has often criticized the idea that markets are unpredictable — and with good reason, since he made his fortune by predicting them. Likewise, wild randomness can be a sign of underlying chaos — which, perhaps counterintuitively, indicates that there is often enough structure present to make useful predictions. And so whatever your views on markets, there's a place for the Predictors. It's no surprise, then, that droves of investors have followed in Farmer's and Packard's pioneering footsteps. In the twenty years since the first computers arrived at the door of 123 Griffin Street, black box models have taken hold on Wall Street. They are the principal tool of the quant hedge funds, from D. E. Shaw to Citadel. The prediction business has become an industry.

CHAPTER 7

Tyranny of the Dragon King

DIDIER SORNETTE LOOKED at the data again. He rubbed his forehead thoughtfully. The pattern was unmistakable. Something was about to happen — something big. He was sure of it, even though predicting such things was notoriously difficult. He leaned back and looked out the window of his office at the University of California, Los Angeles, geophysics institute. Such a tremor could have substantial consequences. The question, though, was what to do about it. Should he issue a warning? Would anyone believe him? And even if anyone did, what could be done?

It was late summer 1997. Sornette had been working on this theory for years now, though the idea of applying it in the present context was new. Still, he had had ample time to test it with historical data. In each instance, before a major event, he had seen this same characteristic pattern. It looked like a wavy line, but with the oscillations getting faster and faster over time, the peaks becoming closer and closer to one another as though they were all trying to bunch up around the same point. The critical point. Sornette had found, both theoretically and experimentally, that these patterns should be robust enough to make predictions, to project when the critical point would occur. The same pattern appeared all over the place: before earthquakes, before avalanches, before certain kinds of materials exploded. But this time it was different. This time, Sornette actually saw the pattern in advance. It was the difference between realizing a prediction was possible — a risk-free endeavor — and actually making it. But Sornette was confident. He would be willing to bet on this.

He picked up the phone and called his colleague Olivier Ledoit. Ledoit was a young faculty member at the Anderson Graduate School of Management at UCLA. Sornette told his friend what he had found. The data showed that a critical event was coming. Earth-shattering, perhaps, but not geological: this event would be a potentially dramatic crash of the world's financial markets. Sornette could even say when it would happen. His calculations put it at the end of October, just a few months away.

Sornette had been working his way into finance for several years, but even so he was still a physicist. Ledoit knew the financial industry and could help him figure out the next steps. The two settled on a plan. First, they would file their warning with the authorities. Sornette and his postdoctoral researcher at UCLA, another geophysicist-cum-economist named Anders Johansen, wrote a notice and sent it to the French patent office. No one would believe them now, of course — none of the traditional methods of analyzing markets pointed to instability. And if they waited until after the crash, no one would believe them either, though for a different reason: their voices would be lost among the thousands of economists and investors who would insist they had seen this coming. The patent filing would be their insurance policy, their proof that they really had made the prediction, over a month before the crash. The notice was filed on September 17, 1997. It predicted a market crash in late October of that same year.

The second step? Profit. It's easy to make money when markets are rising. But in many ways, a market crash is an even more dramatic profit opportunity, if you can see it coming. There are several ways to make money off a crash, but the simplest way is buying put options. The options I discussed earlier are known as call options. You buy the right to purchase a stock at some fixed price, called the strike price, at some time in the future. If the market value of the stock goes above the strike price, you profit, because you have the right to buy the stock at the strike price, and then sell it at the higher market price, pocketing the difference. Of course, if the price doesn't go up, that's OK too. You're only out the money spent on the option, and not the higher price of the stock itself. Put options work in essentially the opposite way. You buy the right to *sell* a stock at a specific price. In this case, you profit if the price of the stock falls below the strike price, because you can buy the stock at the market price and sell it at the higher strike price, again pocketing the difference.

Recall that far-out-of-the-money options are options that will be valuable only if the market takes a dramatic swing. Since dramatic market swings tend to be unlikely, far-out-of-the-money options tend to be very inexpensive (because the people selling them believe they carry little risk). When markets crash, however, these far-out-of-the-money put options can become very valuable indeed, with almost no initial cost. And if you know when the market is going to crash, you can walk away with enormous profits accrued over a very short time — just a few days, say — with relatively little risk. It sure beats buy-and-hold. The problem, of course, is predicting the unpredictable.

Imagine inflating a balloon. You start with a limp piece of rubber. In this uninflated state, the balloon is stretchy and very difficult to tear. You could poke it and prod it any way you like, even with a very sharp knife, and unless you stretch the balloon out first, the knife is unlikely to puncture it. A pin would do no damage at all. Now begin to blow air into it. After a few puffs, the balloon starts to expand. The pressure from the air inside is pushing the walls of the balloon out, just enough to give the surface a roughly spherical shape. The material still has considerable give. Depending on how much air has been pumped in, a very sharp knife might now slice the rubber, but the balloon certainly won't pop, even if you manage to puncture it. A puncture would allow the air inside to leak, but it wouldn't be very dramatic.

As you blow more air into the balloon, however, it becomes increasingly sensitive to outside effects. A fully inflated balloon is liable to pop from the slightest brush with a tree branch or a bit of concrete — a tap from a pin is certain to make it explode. Indeed, if you keep blowing air into a balloon, you can make it burst by touching it with your fingertips, or by simply blowing in another mouthful of air. Once the balloon is primed, it doesn't take much to produce a very dramatic effect: the balloon shreds into tiny pieces faster than the speed of sound.

What makes a balloon pop? In some sense, it's an external cause: a tree branch or a pin, or perhaps the pressure from your fingers as you hold it. But these very same influences, under most circumstances, have little or no effect on the balloon. The balloon needs to be inflated, or even overinflated, for the external cause to take hold. Moreover, the particular external cause doesn't much matter — it's far more important that the balloon be highly inflated when it is pricked. In fact, the external cause of a popped balloon isn't what makes the balloon pop at all. It's the internal instability in the balloon's state that makes it susceptible to an explosive pop.

The bursting of a balloon is one of a variety of phenomena known as ruptures. Ruptures occur in all sorts of materials when they are put under stress. A rupture can often be thought of as a straw-that-broke-the-camel's-back effect: the stress on a substance, such as high internal pressure (caused, for instance, by the air in a balloon, or the gas in a soda can that has been shaken up — or the accumulated weight on a camel's back), leads to instabilities that in turn make the material vulnerable to explosive events. These explosions, sometimes called critical events, are the ruptures. Just as when a balloon bursts, a rupturing material changes its state very rapidly, releasing a substantial amount of energy in the process. Events that might otherwise have little effect, like a pin breaking the surface of an only partially inflated balloon, tend to cascade, building into something larger.

No one has done more to improve our understanding of ruptures than Didier Sornette. He has been stunningly prolific. Still in his early fifties, he has published more than 450 scientific articles in just thirty years. He has also written four books, one on physics, two on finance, and one on Zipf's law, the unusual distribution that first attracted Mandelbrot's attention. But even more remarkable than the amount of work he has produced is its range. Most physicists, even the most successful, work in a handful of closely connected areas. Acquiring expertise in a new area is difficult, and for most people, once or twice is often enough for a lifetime.

But Sornette has made contributions to more than a dozen fields, ranging from material science to geophysics, to decision theory (a branch of economics and psychology), to financial markets, even to neuroscience (he has done considerable work on the origin and prediction of epileptic seizures). He thinks of himself as a scientist in the broadest sense, as someone conversant in the sciences at large. He studied physics as a young man, not because he believed he wanted to devote his life to the field, but because he thought of physics as a kind of mother science. He likes to quote the philosopher Descartes, who in his magnum opus *Discourse on Method* wrote that the sciences are like a tree: metaphysics is the roots, physics is the trunk, and everything else is the branches. (Nowadays, Sornette is more modest about his training. He thinks of his background in physics as an excellent preparation for approaching many problems but says that the intellectual challenges of fields like economics and biology are at least an order of magnitude more difficult than those posed by physics.) Despite the variety of topics, however, much of Sornette's work involves identifying patterns endemic to the structures of complex systems and using these patterns to predict critical phenomena: ruptures, quakes, crashes.

One of Sornette's earliest scientific projects involved ruptures in Kevlar, a synthetic fiber developed in 1965 by Du Pont (and heir to the nylon tradition described earlier). It is a famously strong substance, used in the bulletproof vests worn by police and soldiers, and even as a replacement for steel in suspension bridge cables. It is stronger at very cold temperatures than at room temperature, and it is largely stable at extremely high heat, at least for short periods. It's a marvel of modern chemistry.

These properties have made Kevlar a very attractive material for all sorts of high-tech applications. It was one of these — space flight — that led Sornette to become involved in Kevlar research. Initially, the space race was a two-sided affair, between the United States and the Soviet Union. But by the mid-1960s, the leaders of several Western European nations began to realize that Europe couldn't rely on the largesse of the superpowers to further European economic, military, and scientific interests in space. At first, Europe's entry into the space race was slow and scattered, but then in 1975, the various nascent organizations that had been formed over the previous decade combined into what is now the European Space Agency. By this time, the space race had begun to slow, with further escalation proving too costly for both superpowers. This left an opportunity for the new European agency to rapidly catch up and assert itself as a dominant force in the space industry. A principal part of the new European initiative was a series of cutting-edge rockets called *Ariane,* designed as satellite delivery mechanisms.

In 1983, the still-young European Space Agency began developing a new variety of *Ariane* rocket, the *Ariane 4,* to launch commercial satellites, particularly communication satellites. (It was enormously successful — at one stage, it was used for roughly half of all commercial satellites launched worldwide.) The new rocket was designed by the French space agency, CNES, but manufactured by private contractors. It was one of these private contractors, a firm called Aérospatiale, that contacted Sornette.

Rockets, including the *Ariane,* often require several substances that need to be kept under very high pressure in order for combustion to occur. The chemicals are stored in vessels called pressure tanks — essentially, high-tech water balloons intended to maintain the necessary high pressures without bursting under the strain. The researchers at Aérospatiale who contacted Sornette were studying the behavior of pressure tanks that would be used in the *Ariane 4.* These tanks were made out of Kevlar. Usually, the tanks were strong, even at very high pressures. Except when they suddenly exploded. The group at Aérospatiale was trying to determine the conditions under which this would happen.

We know that if a balloon is inflated sufficiently, it will nearly always pop when pricked with a sharp pin. Other substances, though, can be trickier to figure out. Materials like Kevlar will eventually rupture from the strain of high-pressure contents, but determining precisely when, or why, is a surprisingly difficult problem. When substances like Kevlar are put under significant stress, tiny fractures begin to appear.

Sometimes these fractures combine and grow into slightly larger fractures. Sometimes these slightly larger fractures grow into still-larger fractures, and so on, until you get a very large fracture. These fractures follow a pattern we have already seen: they are fractals, where the tiniest fractures look just like the larger ones. The difficulty is that tiny fractures don't affect the behavior of the pressure tanks, whereas the largest fractures can be disastrous. But it's hard to say what makes a large fracture different from a small one, at least in terms of the fractures' causes. A large fracture is just a small one that never stopped growing; very large, disruptive fractures are no different in kind from the very small benign ones.

This relationship between large and small fractures posed a major problem for the rocket scientists. It meant that even under ordinary working conditions, when the Kevlar was usually stable, there was always a chance that a normal tiny fracture would spontaneously grow into a major one and destroy the rocket. Any given fracture, even the very smallest ones, had the capacity to become explosive. When Sornette joined the team, the other scientists were at a loss. To put these pressure tanks to good use, they needed to figure out how to use them safely — that is, they needed to figure out the conditions under which ruptures would occur. But this seemed an impossible task. The ruptures seemed, quite simply, random.

Until Sornette noticed a pattern.

Normally, the parts of a pressure tank are more or less independent, like workers in the nineteenth century, before collective bargaining. If you kick a pressure tank, for instance, there might be some vibrations, but these will die off pretty quickly, and even if you manage to put a dent in the part of the tank where your foot made contact (unlikely), you won't do any damage to the rest of the tank. Likewise, if a small fracture appears under these circumstances, it won't produce a rupture. This is a bit like when you try to pop an only partially inflated balloon: a pin doesn't have much of an effect.

Sometimes, though, the various parts of the material begin to conspire with one another. They display a kind of herding effect. This can happen for various reasons: heat, say, or pressure, or other external effects. When this occurs, it's almost as if the various parts of the material have unionized. A kick in one place can ripple through a whole tank, with small localized influences leading to dramatic effects, much as a pinprick in one place can make an inflated balloon tear itself apart. This kind of conspiracy is sometimes called self-organization, because no matter how random and uncorrelated the materials are to begin with, if they are placed under stress, they will begin to coordinate their activity. It's as though the bits and pieces of material begin to stir under pressure, gradually deciding to join together in common cause.

Sornette didn't come up with the notion of self-organization, though he has done as much work on the theory as anyone. Instead, he realized something slightly different. He finally understood how a small labor strike differs from a catastrophic one. All strikes are caused by the same sorts of sparks: an egregious injury; an unfair termination; cut wages. You might think that there's no way of telling which such events will lead to a nationwide walkout. A large strike looks like a small strike that, for whatever reason, simply didn't stop. So, too, with the microfractures that, under some circumstances, seem to explode into ruptures that tear a material apart. But the biggest strikes require something more than just a spark: they require a labor movement, with a high degree of structure and a capacity for coordinated action. They require a mechanism for system-wide feedback and amplification, something to transform an otherwise small event into a large one. In other words, if you want to predict a major strike, don't look for the grievances. Those are always there. Look for the unions. Look for telltale patterns of self-organization. Coordination, rather than the pinpricks, is what really leads to critical events. And Sornette would take that insight straight to the bank.

Sornette was born in Paris but raised in the southeast of France, in a town called Draguignan on the French Riviera. Draguignan is about an hour by car from Saint-Tropez, the beautiful Mediterranean resort town famous as a jet-set vacation spot. Through high school, Sornette would often go to Saint-Tropez to sail and wind-surf. Once he graduated, he moved up the coast to Nice where he enrolled in a preparatory school to study for the *grande école* admissions exam. (It was at a similar kind of school in Lyon, a couple of hundred miles north, that Mandelbrot hid from the Nazis during World War II.) Sornette performed extremely well on his exams and was admitted to the most prestigious of the *grandes écoles*, École Normale Supérieure.

He received his doctorate in 1981, at the young age of twenty-four —and was immediately given a tenured position at the University of Nice. His earliest work was in an area of physics known as condensed matter—the study of matter under extreme conditions. But he began to branch out the following year, when he began his obligatory military service. He spent these years working at a government military contractor called Thomson-Sintra (keeping his academic position all the while). It was during this period, working on research for the military, that Sornette first began to study chaos theory and complex systems, subjects that would later provide much of the foundation for his interdisciplinary work.

In June 1986, Sornette married a young geophysicist named Anne Sauron. At the time she was a PhD student in Orléans, interested in geophysics, but after their marriage she moved to Nice, where Sornette was already established. Shortly after their wedding, Sornette secured funding for his new wife to join his research group, with him as her doctoral advisor. They focused on connecting the work Sornette had begun on ruptures to questions concerning the cause of earthquakes.

Although Sornette was officially Sauron's advisor, their work was really a collaboration between experts in different fields. He didn't know the first thing about earthquakes when they began working together (she, meanwhile, didn't know anything about material rupture). But Sornette was a quick study. Together, they began to think about applying fractal geometry to the study of tectonic plates, sections of the Earth's crust that slowly creep around the planet. Tectonic plates were originally proposed to explain the strange evidence that the continents were once connected—for example, certain varieties of plant are found only in western South America and eastern Australia—but they are now believed to be responsible for things like earthquakes (which occur when two plates collide or shift past one another), mountain ranges (which form when the plates collide, buckling at the collision site), volcanoes (which erupt at the interface between plates, where magma from below the crust can escape), and ocean trenches (the opposite of mountain ranges). The Sornettes' work was an attempt to understand how the current geology and topography of the divide between Asia and India — a stretch of land as long across as the continental United States, spanning the Himalayas and a handful of smaller mountain ranges — could arise as a result of many small earthquakes over millions of years, as the two continents collided with one another.

Geophysicists study a broad swath of topics concerning the internal structure of planets. But their bread and butter, the research that gets funding agencies most excited, is predicting natural disasters like earthquakes and volcanoes. Earthquake prediction is a matter of particular importance, for both scientific and humanitarian reasons. It is also famously difficult, though this hasn't stopped scientists, and before that philosophers and astrologers, from trying their hand. The ancient Roman historian Aelian, for instance, hinted that animals could accurately predict earthquakes, claiming that snakes and weasels evacuated the Greek city of Helice a few days in advance of an earthquake that devastated the region. An ancient Indian astrologer and mathematician named Varahamihira believed that earthquakes could be predicted by looking for particular cloud patterns.

In the 1960s and 1970s, the United States and the Soviet Union launched competing earthquake prediction initiatives, showering geophysicists with funds. These programs led to claims that anything from electrical storms to increased radioactivity to an absence of earthquakes could be used to predict future disasters. But the state of the art, especially in the mid-1980s, was not much better than it was when Helice succumbed in 373 B.C. (Indeed, both animal behavior and earthquake clouds remain on the list of active research programs, even today.) The ability to accurately predict earthquakes is a kind of holy grail.

Sornette began his collaboration with Aérospatiale in 1989. That same year, he and Sauron published a paper connecting self-organization, the idea behind the theory of ruptures he had been developing, to earthquakes. The analogy was quite close: the Earth's crust could be understood as a material capable of rupture; a theory that described rupture in something like Kevlar could also, in principle, describe rupture in something like rock. The last step was simply to view catastrophic earthquakes as critical events, ruptures at the interface between tectonic plates. It was not the very first paper to link the ideas of self-organization, criticality, and earthquakes. But it was close. And it set the stage for Sornette to think of his two parallel projects — pressure tanks and earthquakes — as closely connected.

The moment of inspiration came two years later, in 1991. By this time, he and others had developed a detailed model for how fractures and cracks percolate through a material. This model accounted for how degrees of organization and coordination could serve to amplify fractures, to turn small causes into large effects. It was while thinking about this model that Sornette realized that if all of the pieces were in place for a critical event, an explosive rupture, the way in which the fractures leading up to the rupture would multiply would be affected. The idea was that a rupture would be preceded by smaller events, following a very specific, accelerating pattern. This pattern is called log-periodic because the time between the smaller events decreases in a particular way, related to the logarithm of the time. Since this pattern would occur only if the system were primed for a rupture, it counted as a signal that a critical event was about to occur. And because the pattern was one that accelerated over time, if you looked at a few of the smaller events in a row, you could determine whether they were showing the log-periodic behavior (because the time between the events would be shrinking), and you could extrapolate forward in time to figure out when the peaks would collapse into one another, thus predicting the critical event.

Sornette first sought to test the theory with the pressure tanks. Sure enough, right before a rupture, he and his collaborators observed the log-periodic pattern in vibrations of the tanks known as acoustic emissions. Basically, the tanks would start rumbling as fractures began to appear. And if the rumblings were log-periodic, a critical event was about to occur. Aérospatiale quickly patented the method for predicting when its rockets' tanks would explode; it is still used today for forecasting and testing pressure tank failure.

But pressure tanks were only the beginning. If the Sornettes were right about the close connection between material rupture and earthquakes, Didier's discovery had enormous implications. There were all sorts of reasons for the occurrence of *small* earthquakes, which were the equivalent of tiny fractures in Kevlar under stress. But if catastrophic earthquakes were like ruptures, as the Sornettes had proposed, then one should be able to predict a critical earthquake by looking for the log-periodic pattern in the geophysical data. (There is a long history of people believing that small earthquakes foretell larger ones — Sornette's approach makes this much more precise, by saying *when* small earthquakes are predictive.) Sornette's methods weren't useful for predicting anything but the critical earthquakes, the ones that resulted from underlying coordination. But these were usually the biggest earthquakes of all, the ones that leveled cities and tore continents apart. It was a tool for predicting catastrophe. The holy grail, indeed.

As September 1997 crept to an end and October began, Sornette and Ledoit began to buy far-out-of-the-money put options. Neither had a fortune to invest, but the options were cheap. Nervously, they watched as the major world indices marched along, blithely unaware that disaster was just around the bend. Sornette was confident enough to put his own money where his best science told him to. But there have been only a handful of market crashes in modern history. This pattern could have been a false alarm. Much was on the line for Sornette, both financially and intellectually.

The middle of October came and went. Sornette's predictions were not perfectly precise — the market oscillations put the crash somewhere toward the end of October, but it was difficult to pinpoint a specific day. Each day, the probability that the crash would happen (given that it hadn't already) increased. But this would continue for only a short while — it was theoretically possible, if unlikely, for the critical point to pass without so much as a shudder from the markets. Another week passed. Going into the weekend of October 24, there was still no crash. It was becoming nerve-racking. The end of October was here, and Sornette had nothing to show for it.

And then it happened. On Monday, October 27, 1997, the Dow Jones Industrial Average suffered its sixth-largest single-day point loss ever, down 554 points. The NASDAQ and S&P 500 indexes suffered similar losses. For the first time in its history, the New York Stock Exchange was forced to close early in order to avoid a still more severe catastrophe. On that day alone, over $650 billion vanished from New York's financial markets. International markets fared just as poorly, with sharp declines in London, Frankfurt, and Tokyo. The Hong Kong Hang Seng index fell 14% the following night.

Sornette and Ledoit, however, made a 400% profit. They released their Merrill Lynch trading statement that November to prove it. The crash had come, just as Sornette had predicted.

Historians now explain the worldwide crash as a reverberation effect. Earlier that year, the Thai baht collapsed after the Thai government decided to stop pegging it to the U.S. dollar. Thailand carried significant foreign debt before the collapse of the currency, and afterward the country was essentially bankrupt. Thailand's difficulties quickly spread to its neighbors, earning the nickname "Asian flu" for the crisis because of the way it moved through Southeast Asian economies, devaluing currencies and depressing equity markets throughout the region. These conditions increased uncertainty in all parts of the world's economy, leading to unusually high variations in the prices of securities. When Asian markets fell overnight on the twenty-sixth, investors in the United States reacted strongly and amplified the crash.

One of the most striking things about the October 27 crash, and the reason it is now referred to as a "mini crash," is that New York markets rebounded the next day. By the close of trading on the twenty-eighth, the Dow had regained 60% of the previous day's losses. And in a striking counterpoint to closing its doors early for the first time the day before, October 28 was the first day that over a billion shares were traded on the NYSE. This kind of dramatic seesawing is telling: since the cumulative effect of the crash and rebound was a relatively modest change in prices, standard reasoning about pricing in an efficient market does not seem to apply. That is, any theory of the stock market that accounts for price changes in terms of the actual values of the companies whose stocks are being traded would predict that a crash would correspond to some dramatic change in the real-world values. But this didn't happen. Stocks were worth more or less the same amount on October 29 as they had been on October 26, indicating that most investors didn't think the values of the companies had changed all that much. Instead, it seems that the crash resulted from some sort of internal instability in the markets themselves.

According to Sornette and his collaborators, this is a feature that shows up in many market crashes. As he is fond of pointing out, the standard economic reasoning suggests that if bubbles are possible at all, they can end only with some dramatic news that materially changes the value of firms whose stocks are being traded. And yet, many economists agree that if you look at particular crashes, it is often very hard to identify what that piece of news could have been. Sure, there's always *some* piece of bad news to associate with a market crash. But one is often stuck blaming extreme events on run-of-the-mill external causes that do not seem to change the value of the things being traded. This alone should be highly suggestive, at least to someone who is accustomed to thinking about critical phenomena in physics, because it implies that even if a piece of news is the immediate cause of a crash, there is something about the state of the market that determines whether the market actually crashes, or just closes a few points lower. And as with ruptures and earthquakes, Sornette argues, even if you cannot predict the news, you can try to identify when the market is in a precarious state. Just look for the log-periodic tremors.

Critical phenomena often have what physicists call universal properties. This means that you can start with two materials that look as different from one another as can be — a Kevlar tank, for instance, and tectonic plates — and find that, despite the profound differences in their microscopic details, under certain circumstances they exhibit the exact same large-scale behaviors. Both rupture, for instance, as a result of prolonged strain. If you look in detail at how the ruptures occur, you find that the differences in the microscopic details fade away and the radically different materials end up acting in more or less the same way. There are certain universal laws that seem to apply at a statistical level. You might think of these as laws that govern coordination between parts, irrespective of what the parts happen to be. It is this kind of universality that makes Sornette and his collaborators' ideas so widely applicable. The details are often different from field to field, but the principal mechanisms are not. The same phenomena affect avalanches, forest fires, political revolutions, even epileptic seizures.

Sornette's first foray into economics was in 1994. He coauthored a paper with another physicist in France, named Jean-Philippe Bouchaud. That same year, Sornette and Bouchaud went on to found a research company called Science & Finance, which in 2000 merged with a Parisian hedge fund management company, Capital Fund Management (CFM). Today Bouchaud is chairman and chief scientist of CFM, which has grown to be the largest hedge fund management company in France. (He is still officially a physics professor, at École Polytechnique, the *grande école* near Paris where Mandelbrot studied; Sornette, meanwhile, left Science & Finance in 1997.) Their joint paper showed how to price options even if the underlying stock does not follow the kind of random walk assumed by Black and Scholes. This effectively extended the theory of options pricing to more sophisticated models of price changes, including those with fat-tailed distributions. (O'Connor and Associates had already done work along these lines — but this wasn't widely known.)

After that paper, Sornette was hooked. Over the next several years, he read more and more about traditional economics, adding what he could to problems like options pricing and risk. (Sornette prides himself on having learned to think like an economist.) Much of this early work was done in collaboration with Bouchaud, who by this time was working on finance nearly full-time.

In 1996, Sornette's work on earthquakes earned him a part-time professor-in-residence position in UCLA's earth and space sciences department, and at the Institute of Geophysics and Planetary Physics. By this time, though, at least half of his energy was devoted to finance. That same year, Sornette, Bouchaud, and Sornette's postdoctoral researcher, Anders Johansen, realized that Sornette's earlier work on predicting earthquakes and ruptures could be extended to predicting market crashes. They published a paper together in another physics journal. Amazingly, just a few months later, Sornette detected the log-periodic pattern that he had determined should presage a crash. The success of October 1997 deepened his belief that he was on to something important, and he redoubled his efforts on economics and financial modeling.

As with his theories of material rupture and earthquakes, the central idea behind Sornette's market-crash-as-critical-event hypothesis involves collective action, or herding behavior. By itself, this is hardly surprising, as the suggestion that market crashes have something to do with mob psychology is old: in 1841, Charles Mackay wrote a book on, among other things, economic bubbles that he called *Extraordinary Popular Delusions and the Madness of Crowds.* There, he pointed to several historical cases in which entire countries had been taken by some sort of frenzy, leading to speculative bubbles — market conditions under which prices become entirely divorced from the value of the things being traded.

Perhaps the most striking example occurred in the Netherlands in the early seventeenth century. The subject of speculation was tulip bulbs. Tulips originated in Turkey but made their way into western Europe, via Austria, in the middle of the sixteenth century. The flowers were considered very beautiful and were highly prized by the European aristocracy, but the real money was in tulip bulbs, which could be used both to produce the flowers and to produce new bulbs. Tulips came to represent Dutch imperial power. The country's new merchant class, made wealthy by trade in the Dutch East and West Indies, would broadcast its power and prestige with ornate flower gardens, with tulips as the centerpiece.

So tulip bulbs were a valuable commodity. But how valuable? During the 1630s, prices began to grow rapidly. By 1635, trades worth 2,500 Dutch guilders (worth roughly $30,000 in 2010 dollars) for a single bulb were recorded. Trades of 1,500 guilders were common. In contrast, a skilled laborer could expect to make about 150 guilders in a year. Around this time, foreign money began to pour into the market as outsiders tried to make a quick buck in the tulip game. The Dutch were thrilled. They took the foreign investment to mean that all of Europe was catching on to their tulip craze, and so they doubled down: ordinary people sold their belongings, mortgaged their houses, and exhausted their savings to participate in the tulip market.

Tulips bulbs are typically planted in the fall and then harvested in the late spring. But winter was the prime time for speculation because this was when would-be investors had the least information about the supply for the coming year: the old bulbs had been planted but the new bulbs and cut flowers were not yet available. It was during the winter of 1636–37 that tulip mania (as it is now called) reached its height. That winter, a single bulb sold for as much as 5,200 guilders (more than $60,000 for one tulip bulb!). And then one day in February 1637, at an otherwise ordinary tulip auction in Haarlem, the bidding stopped too soon. Apparently no one had invited the next batch of tulip fools. That day, prized tulips sold for just a fraction of what they had even one day before. Panic spread quickly, and within days prices had fallen to less than 1% of their height. Fortunes that had been made overnight vanished by morning. The Dutch economy teetered, until ultimately the government needed to intervene.

Herding and similar phenomena — the kinds of behavior that lead to bubbles — seem to be an ever-present aspect of human psychology. No one wants to be left out, and so we tend to copy one another. Ordinarily, though, we do not act like lemmings. Even if we look to one another for guidance, we do not usually follow blindly. The question, then, is why under some circumstances herding seems to take over. How does something like tulip mania strike? When do the normal mental brakes that would keep someone from spending his entire life savings on a tulip bulb give out? Sornette doesn't have an answer to this question, though he has developed some models that predict which circumstances will lead herding effects to become particularly strong. What Sornette *can* do is identify when herding effects *have* taken over. This amounts to identifying when a speculative bubble has taken hold in a particular market and to predicting the probability that the bubble will pop before a certain fixed time (the critical point).

Despite Sornette's enormous productivity in finance, he resists the idea that he has "switched over" to economics. Since 2006, he has held the Chair of Entrepreneurial Risks at the Swiss Federal Institute of Technology in Zürich (usually abbreviated ETH Zürich) — his first finance-related academic position — but he maintains a part-time position in geophysics at UCLA, and also a full-time appointment as a geophysicist in ETH Zürich's physics department. He continues to write articles and supervise students in both fields. And if you ask him what prompted the change in focus of his work, since surely there was a shift in the mid-1990s when he began working on new topics, he replies, with some bewilderment, that he has always been interested in such things. After all, he is interested in everything.

Still, he does think there is something special about finance and economics. Many people go into science because of some urge to understand how the world works. But, Sornette believes, the physical world is only part of the story. He is just as interested, perhaps more interested, in how the social world works. Gravity may keep the planet in orbit, but, as the emcee in the musical *Cabaret* sings, money makes the world go round. And financial markets determine how money flows. As Sornette puts it, finance is the "queen, and not the maid." It controls everything. And whatever your political position on the role of financial markets in global geopolitics, Sornette believes that the very fact that financial markets and the people who run them *do* have so much social power is a sufficient reason to look closely at how they work.

Since first predicting the October 1997 crash, Sornette has had a remarkable track record of identifying when market crashes will occur. He saw the log-periodic pattern in advance of the September 2008 crash, for instance, and was able to predict the timing. Similarly, the 1998 collapse in the Russian ruble that brought Long-Term Capital Management to its knees showed the signs of an impending crash — indeed, Sornette has claimed that even though the largely unanticipated Russian debt default may have triggered the market turmoil that summer, the crash showed the log-periodic precursors characteristic of herding behavior. This means that a market crash would likely have occurred during that period whether the ruble had collapsed or not. The balloon was already in a primed state; Russia's default was just the pinprick.

He has had success predicting other crashes, as well, most notably the dot-com crash that occurred in 2000. Over several years in the late nineties, technology stocks skyrocketed. In 1998 and 1999, the technology sector of the S&P 500 index went up by a factor of four, while the index as a whole increased by just 50%. The technology-based NASDAQ index increased by almost a factor of three between 1998 and early 2000. Analysts began talking about a so-called new economy consisting of computer firms and companies whose business strategies depended entirely on the Internet. For these companies, none of the old rules applied. It didn't matter if a firm was making any money, for instance — earnings could be negative, but the company could still be considered valuable if there was a wide expectation of success in the future. In many ways, the boom echoed earlier periods of speculation: in the 1920s, for instance, investors also spoke of a "new economy," though then the hot tech companies were AT&T and General Electric.

Sornette started seeing the log-periodic oscillations in NASDAQ data beginning in late 1999. By March 10, 2000 — the day the NASDAQ peaked — he had enough data to say the crash was imminent, and to predict when it would occur. He put the date somewhere between March 31 and May 2. Sure enough, during the week beginning April 10, the NASDAQ fell by 25%. Tech stocks had gone the way of the tulip bulb.

The methods Sornette has used to identify bubbles and predict when crashes will occur can also be used to identify a situation that Sornette has called an anti-bubble. These are cases in which stock prices are artificially low. On January 25, 1999, for instance, Sornette posted a paper on an online physics archive claiming that, based on his observation of log-periodic patterns in the market data, the Japanese Nikkei stock index was in the midst of an anti-bubble. The paper included quite precise predictions: Sornette indicated that by the end of that year, the Nikkei would increase by 50%.

This prediction was all the more remarkable because the Japanese market was near its fourteen-year low, which it reached on January 5, 1999. All indications were that the market would continue to fall — an opinion held by most economists at the time. Nobel Prize laureate and *New York Times* opinion columnist Paul Krugman, for instance, wrote on January 20 that the Japanese economy was beginning to look like a tragedy, and that there simply wasn't enough demand for a recovery. But time proved Sornette right. By the end of the year, the Nikkei had recovered, by precisely the 50% Sornette predicted.

Mandelbrot's work gave some economists reason to think that markets are wildly random, exhibiting behavior that someone like Bachelier or Osborne could never have imagined. Even if Mandelbrot turned out to be wrong in the details of his proposal, he nonetheless revealed that financial markets are governed by fat-tailed distributions. There's nothing special about extreme financial events. They are not exceptions; they are the norm — and worse, they happen all the time, for the same reason as more mundane events. Big market drawdowns, at their core, are just smaller drawdowns that didn't stop.

If this is right, one might think that there is no way to predict catastrophes. Indeed, self-organization, one of the principal parts of the theory of critical phenomena, is usually associated with just the kind of fat-tailed distributions that make predicting extreme events so difficult. The three physicists who first introduced the notion of self-organization, Per Bak, Chao Tang, and Kurt Wiesenfeld, took their discovery as evidence that extreme events are, in principle, indistinguishable from more moderate events. The moral, they thought, was that predicting such events was a hopeless endeavor.

This concern is at the heart of hedge fund manager Nassim Taleb's argument against modeling in finance. In his book *The Black Swan,* Taleb explains that some events — he calls them "black swans" — are so far from standard, normal distribution expectations that you cannot even make sense of questions about their likelihood. They are essentially unpredictable, and yet when they occur, they change everything. Taleb takes it to be a consequence of Mandelbrot's arguments that these kinds of extreme events, the events with the most dramatic consequences, occur much more frequently than any model can account for. To trust a mathematical model in a wildly random system like a financial market is foolish, then, because the models exclude the most important phenomena: the catastrophic crashes.

Recently, Sornette introduced a new term for extreme events. Instead of black swans, he calls them "dragon kings." He used the word *king* because, if you try to match plots like Pareto's law—the fat-tailed distribution governing income disparity that Mandelbrot studied at IBM—to countries that have a monarchy, you find that kings don't fit with the 80–20 rule. Kings control far more wealth than they ought to, even by the standards of fat tails. They are true outliers. And they, not the extremely wealthy just below them, are the ones who really exert control. The word *dragon*, meanwhile, is supposed to capture the fact that these kinds of events don't have a natural place in the normal bestiary. They're unlike anything else. Many large earthquakes are little ones that, for whatever reason, didn't stop. These are not predictable using Sornette's methods. But dragon-king earthquakes, the critical events, seem to require more. Like ruptures, they happen only if all sorts of things fall into place in just the right way. A good example of a dragon king is the city of Paris. France's cities follow Zipf's law remarkably well. The distribution of cities in France is fat-tailed, in that the very biggest cities are much bigger than the next biggest cities. But if you plot the size of French cities by their population size, as Zipf's law would have you do, Paris is still much too big. It breaks the mold.

Taleb's argument trades on the fact that black swans can have enormous consequences. Dragon kings are similar in their influence. They are tyrannical when they appear. But unlike black swans, you can hear them coming. Sornette does not argue that all black swans are really dragon kings in disguise, or even that all market crashes are predictable. But he does argue that many things that might seem like black swans really do issue warnings. In many cases, these warnings take the form of log-periodic precursors, oscillations in some form of data that occur only when the system is in the special state where a massive catastrophe can occur. These precursors arise only when the right combination of positive feedback and amplifying processes is in place, along with the self-organization necessary to make a bang, and not a whimper.

The Prediction Company, on the one hand, and Sornette, on the other, offer two ways in which one might fill in the gaps in the now-standard Black-Scholes-style reasoning. The Prediction Company's methods might be thought of as local, in the sense that their strategy involved probing the fine-grained financial data produced every instant by the world's markets for patterns that had some temporary predictive power. These patterns allowed them to build models that could be used over a short window of time to make profitable trades, even though the patterns were often fleeting. Along with these methods, they developed the tools necessary to evaluate the effectiveness of the patterns they were finding, and to tell when they had passed their prime. In a way, the Prediction Company's approach is modest and conservative. It is easy to see why it should work, as a part of what makes markets more efficient.

Sornette, conversely, has taken a more global approach, looking for regularities that are associated with the biggest events, the most damaging catastrophes, and trying to use those regularities to make predictions. His starting point is Mandelbrot's observation that extreme events occur more often than a normal random walk would predict; Sornette believes that catastrophic crashes happen *even more* than Mandelbrot proposed. In other words, even after you accept fat-tailed distributions, you still see extreme events unusually often. Sornette's intuition, on seeing these apparent outliers, is that there must be some mechanism that, at least sometimes, amplifies the largest catastrophes. This is a riskier hypothesis — but it is one that can be tested, and so far, it seems to have passed.

If you think of Mandelbrot's work as a revision to the early accounts of random markets, pointing out why they fail and how, then Sornette's proposal is a second revision. It is a way of saying that, even if markets are wildly random and extreme events occur all the time, at least *some* extreme events can be anticipated if you know what to look for. These dragon kings can upend the entire world economy — and yet they can be studied and understood. They are the stuff of myths, but not of mystery.

CHAPTER 8

A New Manhattan Project

ANOTHER DEBATE. Pia Malaney put her arms on the table and leaned in to listen to her fiancé, Eric Weinstein. Weinstein was a postdoctoral researcher at MIT who had recently finished a PhD in mathematics at Harvard. They were sitting in a bar in Cambridge, Massachusetts, where Weinstein was holding forth on how the ideas used in *his* dissertation could be applied to *hers*. The trouble was that his work had been on an application of abstract geometry to mathematical physics. Her work, meanwhile, was in economics. The two projects seemed as different as could be. She sighed as she recalled, with a sense of the irony, how much easier these discussions had been before she had won him over to her side.

Malaney had met Weinstein in 1988, while he was a graduate student and she was an undergraduate economics major at Wellesley, the women's college located just outside of Boston. Back then, Weinstein had a dim view of economics — a view shared by many of his mathematician colleagues. He thought it consisted of mathematically simple theories that couldn't hope to capture the full complexity of human behavior. Weinstein would get a rise out of friends in the economics department by calling their field "cocktail party conversation": unsubstantial, trivial. He would happily have admitted that he didn't know much about economics, because, after all, there wasn't much to know.

Malaney was not fond of the view frequently espoused by her fiancé. For years, she steadfastly defended her colleagues' work against Weinstein's attacks.

And then one day, she found she had convinced him. All of a sudden, he went from trying to tell her that economics was worthless to declaring that they should collaborate. All Weinstein could talk about was how, with his training in mathematics and physics and her training in economics, they could tackle all sorts of problems that had stumped economists in the past. The point had long been to get her boyfriend to read enough economics to understand that there was substance behind it. Now, though, Malaney found herself wading into the world of mathematical physics. It was not what she had bargained for.

Still, she couldn't deny that their collaboration was already proving fruitful. They had begun to focus on something called the index number problem. The problem concerns how to take complex information about the world, such as information about the cost and quality of various goods, and turn it into a single number that can be used to compare, say, a country's economic health and status at one time to its economic status at another time. Some familiar examples are market indices like the Dow Jones Industrial Average or the S&P 500. These are numbers that are supposed to encode all of the complicated information about the state of the U.S. stock market. Another index that one often hears about is the Consumer Price Index (CPI), which is supposed to be a number that captures information about the cost of the ordinary things that a person living in a U.S. city buys, such as food and housing. Index numbers are crucially important for economic policy because they provide a standard to compare economic indicators over time, and from place to place. (The *Economist* magazine has proposed a particularly straightforward index, called the Big Mac Index. The idea is that the value of a Big Mac hamburger from McDonald's is a reliable constant that can be used to compare the value of money in different countries and at different times.)

Together, Malaney and Weinstein developed an entirely novel way of solving the index number problem by adapting a tool from mathematical physics known as gauge theory. (The early mathematical development of modern gauge theory—the topic on which Weinstein wrote his dissertation—was largely the work of Jim Simons, the mathematical physicist turned hedge fund manager who founded Renaissance Technologies in the 1980s.) Gauge theories use geometry to compare apparently incomparable physical quantities. This, Malaney and Weinstein argued, was precisely what was at issue in the index number problem—although there, instead of incomparable physical quantities, one was trying to compare different economic variables.

It was an unusual, highly technical way of thinking about economics. This made Malaney a little nervous, since she didn't know how economists unaccustomed to such high-level mathematical analysis would react. But she decided to pursue the project for her dissertation after she showed it to her advisor, a superstar in the Harvard economics department named Eric Maskin. (He would go on to win the 2007 Nobel Prize in economics, for work he had already done before meeting Malaney.) Maskin told her the idea was great. He believed she'd made real progress on an important topic, one with long-term political and economic implications. She finished the dissertation during the summer of 1996 and began to think about applying for tenure-track jobs at top research universities. With such a groundbreaking thesis topic and the support of her advisor, she had every reason to think she'd be a competitive candidate for these highly desirable positions. She was living the academic dream.

How much is money worth? This might seem like an odd question. For most people, money doesn't have intrinsic value. The value of money comes from what you can do with it. Perhaps money can't buy you love, but it sure can buy you orange juice, or a pair of pants, or a new car. And over time, the amount of money it takes to buy that same orange juice, pair of pants, or new car changes. Usually, goods become more expensive over time (at least if you look at the price tags alone); grandparents the world over will tell you how little a chocolate bar used to cost, or a movie ticket. A nickel, we're told, went a lot farther in 1950 than it does now. This decrease in the value of money over time is what we usually call inflation.

But how do you measure inflation? It's not as though all prices go up evenly across the board. Even as some goods have become more expensive with time, others have become cheaper. Consider that the price tag for an Apple II, one of the first mass-produced personal computers, with a breakneck processor speed of 1MHz and a whopping 48KB of memory, was $2,638 when it first went on sale in 1977. Nowadays, almost thirty-five years later, you can get a desktop computer with a processor over three thousand times as fast, and with a hundred thousand times more memory, for a fraction of that — just a few hundred dollars. So what if chocolate is more expensive: computing power is now dirt cheap by 1970s standards.

One way in which economists deal with this problem is by looking at how prices change across a broad range of products. They do this by tracking the price of what is called a standard market basket: an imaginary shopping cart filled with groceries and household commodities like gasoline and heating oil, as well as services like education, medical care, and housing. This is what's used to calculate the CPI, which is effectively the average price of the various goods and services in the cart. By looking at price changes for many different items in this way, you can get a rough estimate of how far a dollar (or a euro, or a yen) goes today, as compared to sometime in the past. Gasoline prices might spike over the course of a few months, while computer prices might drop gradually over a few years, but the change in the standard market basket is supposed to be a relatively stable indication of how much spending power changes with time.

Given the role that the CPI plays in calculating things like inflation, it's important to get it right. Unfortunately, this is a difficult thing to do. For one, what should go into the market basket? People with different lifestyles often spend their money very differently: a family with children living in upstate New York buys very different things (for instance, winter coats and heating oil) from a single man living in Southern California (surfboards?); farmers in Iowa have different needs and preferences from coal workers in West Virginia. It is hard to see how a single market basket could reflect the full variation of these different lifestyles. For this reason, the U.S. Bureau of Labor Statistics, which calculates the CPI in America, actually produces many different indices, corresponding to people working in different industries, living in different areas, and so forth.

But this kind of variability hints at a deeper problem. If the things that a person or family buys can vary from family to family, or from place to place, these kinds of preferences can presumably vary with time, too. This can happen on large scales and small scales. Imagine the standard market basket from 1950, long before cell phones or personal computers, when relatively few people went to college or took an airplane on a family vacation. If you looked at the present prices of that standard market basket, you would not have a very good indication of today's cost of living. But so too if you looked at the kinds of things on which someone spent money over a relatively short period of time: say, a standard market basket for someone immediately out of college, and the basket for someone a few years later, after settling down and getting married, or a few years later still, after having kids. Changes in culture, demographics, and technology can all compound to make assigning a number to inflation, or to changes in the cost of living, seem impossible. This is what makes the index number problem so difficult: you need a way to compare values at different times, and for people living very different lifestyles.

The CPI is a blunt tool. Virtually everyone in economics agrees that we need to find some way to hone it. Still, it is incredibly important for policymaking because of its central role in determining inflation, which in turn affects virtually every aspect of the budget. In the United States, for instance, the thresholds for tax brackets are tied to the stated rate of inflation. So are wage increases for government employees. Social Security outlays are also determined by inflation. Every year, these quantities are recalculated based on the inflation rate of the previous year, to adjust for changes in the cost of living. In June 1995 the U.S. Senate appointed the Advisory Commission to Study the Consumer Price Index, usually called the Boskin Commission after Michael Boskin, the Stanford economics professor who chaired it. The brainchild of soon-to-be-disgraced Senator Bob Packwood, then chairman of the Senate Finance Committee, the Boskin Commission was charged with coming up with a better way to compute the CPI, and by extension inflation.

For Malaney and Weinstein, the Boskin Commission seemed like a godsend. A Senate-appointed committee tasked with solving just the problem that they had chosen to tackle made Malaney and Weinstein's work immediately relevant. It was the perfect opportunity for them to make a contribution — not just to economic theory, but potentially to public policy, since Packwood planned to implement the Boskin Commission's findings immediately. Even better, one of the economists appointed to the commission, Dale Jorgenson, was a member of the Harvard economics department.

Hermann Weyl was offered the position of chair of the mathematics department at ETH Zürich (the school where Didier Sornette currently teaches) in 1913, when he was just twenty-seven years old. He arrived in Zürich from Göttingen, a German university that in the early 1920s represented the very pinnacle of international mathematics. His advisor there, David Hilbert, was widely recognized as the most influential mathematician of his day. As Hilbert's student at Göttingen, Weyl was at the center of the mathematical world.

Things were different in Zürich. ETH Zürich had a fine reputation, but it was quite new: it was only in 1911 that ETH was restructured to become a real university, with graduate students, shedding its past as an engineering-oriented teaching school. The other university in the city, the University of Zürich, was the largest in Switzerland. But it was no Göttingen.

Weyl wasn't ETH's only recent hire, however. As part of the restructuring, the school had made a number of appointments to the physics department. One of these was a prominent young physicist, an undergraduate alumnus of ETH named Albert Einstein. Einstein had gone on to do a PhD in physics at the University of Zürich, graduating in 1905 — the same year that he published a mathematical treatment of Brownian motion (anticipated, of course, by Bachelier), came up with a theory of the photoelectric effect (for which he would win the Nobel Prize in 1921), and discovered the special theory of relativity, including his famous equation $e = mc^2$. And yet, none of this led to much success for Einstein. After finishing graduate school, he moved about 150km away to Bern, where the only job he could find was as a patent clerk. Occasionally, he was permitted to teach at the local university.

Gradually, however, as more physicists came to understand the importance of the 1905 papers, Einstein's reputation grew. In 1911, he was offered a professorship at the German university in Prague; the next year, his alma mater offered him a job. By the time Einstein returned to Zürich, he was already a shining star of the physics community. His reputation had exploded in just a few years. He didn't stay at Zürich for long — in 1914, he was appointed director of the Kaiser Wilhelm Institute in Berlin — but the year that Einstein and Weyl spent together was enough to change the course of Weyl's research. Though initially a mathematician in the purest sense, Weyl found Einstein's relativity theory captivating, particularly because when they met, Einstein was just beginning to realize the importance of high-powered modern geometry to the theory.

The basic idea underlying general relativity is that matter — ordinary stuff like cars and people and stars — affects the geometrical properties of space and time. This geometry, meanwhile, determines how bodies move. It is this movement of massive objects through deformed space and time that we ordinarily think of as gravitation, the physical phenomenon that keeps us firmly planted on the surface of the Earth, and that keeps the Earth in its elliptical orbit around the sun. The general relativistic picture is as different as can be from the older, Newtonian theory of gravity. In Newtonian gravitation, space and time are static. Their properties are unrelated to the matter that's distributed through space. Bodies gravitate toward one another via an unexplained force that acts instantaneously at a distance.

Matter affects space and time in Einstein's theory by inducing curvature. When physicists and mathematicians say something is "curved," they mean just what we would ordinarily mean. A tabletop or an unfolded piece of paper is flat; a basketball or a paper towel roll is curved. But from a mathematical point of view, the thing that distinguishes a tabletop from a basketball isn't that a basketball rolls and a table doesn't, or that it's easier to stand on a table than on a basketball. Instead, the feature that characterizes curvature for a mathematician is how hard it is to keep an arrow pointing in the same direction as you move it around the surface. If an object is flat, it turns out to be very easy. Not so if the object is curved.

I admit that this is a weird thing to say. But it isn't hard to see how it works in practice. First, imagine you're standing on a city sidewalk, somewhere in midtown Manhattan, say, where the streets are laid out like a grid. Try to picture what would happen if you did a clockwise lap around the block, all the while trying to keep yourself pointed in one direction — north, say, toward the Bronx. (The direction you're facing, here, is taking the place of an arrow.) You might begin by walking forward for a while as you head uptown. When you get to the next corner, you would head right, east on the crosstown street. But you aren't allowed to turn your body at the corner, since you're trying to stay pointed in the same direction all the time. This means you have to walk sideways down the cross street. And when you get to the next corner, where you should start heading south again, you have to walk backward. If you follow these instructions, never once turning your body as you do the lap, you should find yourself back at the original corner looking in just the same direction as before.

This might not come as a surprise. After all, you never turned your body — why in the world wouldn't you be facing in the same direction? But now let's imagine a longer journey. Instead of doing a lap around the block, imagine trying to keep yourself pointed in the same direction — it might as well be north — as you circumnavigate the globe. For the first leg of your trip, you're going to start in New York and just head east, toward Europe. When you arrive in France, you're going to start crab-walking your way toward Asia, all the while keeping your face firmly pointed toward the North Pole. After a very long (and probably uncomfortable) walk, you will finally reach the Pacific Ocean, and then you'll head for California. When you finally arrive in New York, if you never turned your body, you should still be facing north.

Here's a different itinerary that begins and ends in the same place. You start by heading east, just as before. When you get to Kazakhstan, though, you take a detour. Instead of continuing on toward China, you strike north into Russia. (Now, at least, you get to walk forward.) You head all the way to the Arctic Circle, without turning your body. When you reach the North Pole, you see that New York is directly in front of you, far to the south. You keep moving forward into northern Canada, and then work your way down the Hudson until you return to New York. But this time, when you return to the place where you began, you're facing a different direction: due south! What's gone wrong? You didn't turn your body at any point of the journey, and yet at the end you're facing the opposite way from the direction you started facing — and from the direction you were facing at the end of your first journey.

The reason you end up facing in a different direction after your second round-the-world trip is that the globe is a curved surface (see Figure 5). A city block, meanwhile, is flat. (At least to a first approximation — real city blocks lie on the surface of the Earth, which of course is curved. But you don't see the effects of this curvature over short distances.) If you imagine an ant trying to perform the same experiment on a kitchen table, you would find that, no matter what route the ant took, it would always end up facing in the same direction. This is what a mathematician means when he says that a surface, or a shape, is flat: it exhibits "path independence of parallel transport" (parallel transport because the goal is to try to keep your body parallel to its last direction at all times). For curved surfaces, meanwhile, the direction an arrow points at the end of a journey is "path dependent." On a curved surface, different routes can lead to different results.

The connection between path dependence and curvature may be unfamiliar to non-mathematicians. But the basic idea of path dependence isn't. It is easy to find examples in day-to-day life of things that are path dependent, and things that are path independent. If you go to the store and buy groceries, the amount of milk you have when you get back home is path independent. The amount of milk isn't going to change if you take a different route home from the store. The amount of gasoline in your tank, however, is path dependent. If you take the direct route home, you will usually have more gasoline left when you arrive than if you had taken the scenic route. Path dependence of par-

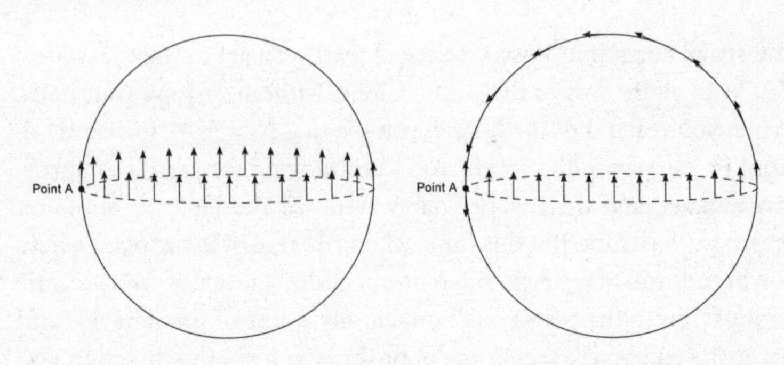

Figure 5: If you move an arrow along a path on a curved surface, being careful to keep the arrow pointing in the same direction at all times, the direction that the arrow points at the end of the path will depend on the path taken. Mathematicians call this property of curved surfaces "path dependence of parallel transport." In this figure, there are two paths around a sphere. The first path takes the arrow from point A around the equator and back to point A. At the end of the trip, the arrow faces the same direction as when it began. The second trip again starts at point A and travels around the equator, but only halfway. On the other side of the sphere, the path moves up over the North Pole and returns to point A that way. At the end of *this* trip, the arrow is pointing in the *opposite* direction from when the trip began. Weyl observed that it was possible to construct physical theories in which not only was *direction* path dependent, but so was the *length* of an arrow. The physical world doesn't actually work that way, but in the years since Weyl first came up with his theory — which he called a gauge theory — many physicists and mathematicians have adapted the mathematics he invented to other problems, with much more success.

allel transport is just a special case of the more general fact that sometimes, things depend not just on where you start and where you end up, but also on the road you take to get there.

Einstein's theory of general relativity makes essential use of the fact that space and time are curved in the sense that parallel transport is path dependent. But Weyl thought that Einstein hadn't gone far enough. In general relativity, if you begin with an arrow at one place and then move it around a path that brings it back to the starting point, it might face a different direction. But it will always have the same length. Weyl thought this was an arbitrary distinction that couldn't have physical meaning, and so he came up with an alternative theory in which length, too, was path dependent, so that if you moved a ruler around two different closed paths, it would have different lengths when it returned to the starting point, depending on the path it took.

Weyl called his new theory a gauge theory. It was the first time the term had been used, and it was based on the idea that there was no universal, once-and-for-all way to "gauge," or measure, the length of a ruler. Suppose you and your neighbor are both about to leave your driveway in the morning on the way to work. Imagine you drive identical cars, and you both work at the same location. What would you say if someone stopped you and asked which car would have more gasoline in the tank when you both got to work, yours or your neighbor's? You might glance at your gas gauge and see that you have a full tank, and then ask your neighbor how much gas he has. But this isn't enough information to answer the question. The answer will depend on the paths you and your neighbor take to work: you might take a direct route, while the neighbor takes the scenic route. Your neighbor might take a highway, while you stick to city streets. Whatever the case may be, how much gasoline each of you has left at the end of your journeys will depend on the paths you take to work. Comparing some path-dependent quantities does not yield a straightforward answer.

This was the sense in which, in Weyl's theory, there was no universal way to measure a ruler, since there was no path-independent way to compare two rulers in different locations. But Weyl realized that this wasn't necessarily a problem: if you wanted to compare the length of a ruler in Chicago to the length of a ruler in Copenhagen, or on Mars, all you needed to do was figure out a way to bring the rulers to the same place so you could hold them up next to each other. This wouldn't be path independent, but that was OK, as long as you could figure out how the change in length would depend on the path you took. In other words, Weyl realized that what really mattered to his theory was identifying a mathematical standard by which comparisons of length could be made—a way of "connecting" different points in a principled way, so that you could compare rulers, even though length was path dependent. Mathematically, what Weyl accomplished was to show how to compare two otherwise incomparable quantities, by moving them to a common location where their properties (in this case, their lengths) could be compared directly.

Weyl's theory wasn't a success. Einstein quickly pointed out that it was inconsistent with some well-known experimental results, and it was soon relegated to the dustbins of scientific history. But Weyl's basic idea about gauge — that to determine if two quantities are equal in a physical theory, you need a standard of comparison that accounts for possible path dependence — was destined to be far more important than the theory that led to it. Gauge theory was resurrected in the 1950s by a pair of young researchers at Brookhaven National Laboratory named C. N. Yang and Robert Mills. Yang and Mills took Weyl's theory one step further: If it was possible to construct a theory in which length was path dependent, was it possible to construct theories in which still other quantities were path dependent? The answer, they realized, was yes. They went on to develop a general framework for much more complicated gauge theories than the one Weyl had imagined.

These theories, now known as Yang-Mills theories, spawned what is sometimes called the gauge revolution. Beginning in 1961, fundamental physics was rewritten in terms of gauge theory — a process that only accelerated when Yang, in collaboration with Jim Simons of Renaissance, realized a deep connection between Yang-Mills gauge theories and modern geometry later that decade. Gauge theories proved particularly important in physics because they proved to be a natural setting to look for "unified" theories, where what was being unified was the standard by which different quantities were compared in the theories. By 1973, it appeared that the three fundamental forces of particle physics — electromagnetism, the weak force, and the strong force — had been unified into a single gauge-theoretic framework. This framework was called the Standard Model of particle physics. Today, it is the single best-confirmed theory ever discovered, in any field. It is the very heart of modern physics.

Jobs in academia, especially the most desirable jobs as tenure-track professors, work on a fixed schedule. Toward the end of each summer, students who are close to finishing their dissertations decide whether they are going to apply that year. If the student and his advisor decide that the dissertation is far enough along, the student begins to put together a dossier, including letters of support from faculty members, examples of the work that will go into the student's dissertation, and a statement describing the student's research interests. Then, come fall, departments that are looking to hire new faculty members advertise their open positions, with applications due at the end of November. If you're lucky, you will be interviewed by a hiring department, and if that goes well, you will be flown out to visit the schools that are interested in you, to give a talk on your dissertation research. In many disciplines, including economics, the process is called "going on the market," an apt phrase for what is essentially an academic cattle call. It's an extraordinarily stressful process. More than anything else, an academic's success on the market is what determines the trajectory of his career.

A graduate student's research history and the quality of his dissertation are crucial in determining whether he will get an academic job. But more important than either of these factors is the strength of the letters written in support of the student by the faculty. If famous, well-respected professors say your research is good or important, that can make all the difference. Each year, the Harvard economics department holds a faculty-wide meeting to determine which of that year's batch of students are going to get the full-throated support of the university's eminent economics faculty. The department goes through each of the candidates, and the student's advisor brings the rest of the department up to speed on the student's research and prospects. It's a closed-door affair, and so only faculty members in the department know exactly what happens. But at the end of the meeting, some students emerge with the wind at their backs. When hiring departments start calling, these students get special endorsements. Others aren't so lucky.

Given the importance of her work, and the strong support of her advisor, Pia Malaney had every reason to expect that she would fare well in this process. Everything was in place. But then came the October jobs meeting. Afterward, she and Maskin met to discuss her job prospects, in light of the department's determination. Things no longer looked so good.

Going into the meeting, Maskin was convinced that her thesis was terrific. But not everyone in the department agreed. One person in particular had reservations: Dale Jorgenson, one of Harvard's two representatives on the Boskin Commission and an expert on the index number problem. Malaney's project covered exactly the same ground that the Boskin Commission was supposed to investigate. She had developed an elegant mathematical framework for addressing precisely the problem they were tasked with. And so, when she learned of his appointment, Malaney arranged a meeting with him. Excited, she described her work to him, showing how gauge theory could be applied to this important problem. Jorgenson replied by throwing her out of his office. "You have nothing," he told her.

At the time, Malaney was discouraged, but she didn't give up. So what if she couldn't convince Jorgenson on her first try? Maskin liked the ideas and would advise the thesis. In the long run, the work would speak for itself. But then, as Malaney prepared to apply for jobs, this vision of the future began to dissolve. During the jobs meeting, it became clear that Jorgenson's resistance to Malaney's project ran deep. Several months later, when the Boskin Commission released its findings, the reasons for his resistance would become clear.

It took years for Malaney to convince Weinstein to take economics seriously. She tried everything: pointing to famous economists, explaining their most influential theories, describing important experimental results. But Weinstein was resistant. The mathematics, he was convinced, was too simple; the subject matter, too complex. Economics was a worthless pursuit, a pseudoscience. Finally, on the verge of giving up, Malaney tried one last tack. She gave Weinstein a challenge, a problem whose solution was equivalent to a classic result in economics known as Coase's theorem.

Ronald Coase was a British economist who spent most of his career in the United States, at the University of Chicago. He was interested in something he called "social cost." Imagine you are the local sheriff in an agricultural community. Two of your constituents come to you, asking you to help them settle an ongoing dispute. One of them is a rancher, raising cattle. The other, the rancher's neighbor, farms soybeans. The dispute concerns the rancher's cattle, which have a habit of wandering over to the farmer's land and destroying his crop. Matters have recently become especially difficult because the farmer has learned that the rancher wants to add more cattle to his herd, and the farmer is concerned that the problem will get worse. What should you do?

When Coase tried to formalize an answer to social cost problems like this one, he came to a striking conclusion. It doesn't matter what the sheriff does, at least from a long-term perspective, as long as three conditions are met: the damages involved must be adequately quantified, some well-defined notion of property must be instituted, and bargaining must be free. To see why this would be, consider what would happen if the sheriff told the rancher he could have as many cattle as he liked, but that he had to pay for all of the damage his herd inflicted. In essence, the rancher has incurred an additional cost to raising cows. Depending on how much damage gets done, and how much soybeans are worth, it may well make sense for the rancher to keep adding head to his herd even while paying the farmer for the soybeans that keep getting destroyed. If the rancher really is paying for the value of the soybeans, the farmer shouldn't care whether the revenue comes from selling the soybeans himself or from the rancher's compensation — in fact, he might as well think of the rancher as a customer buying whatever soybeans the cattle destroy. Ultimately, the rancher and farmer will reach an agreement about how many cattle the rancher will own based on what is maximally profitable for both parties. But what if the sheriff makes some other choice? If the farmer has to pay the rancher to keep his cattle from destroying the farmer's crops, one would expect the exact same bargaining to occur. Coase's theorem says that the endpoint will always be the same: both parties will agree on an arrangement that is maximally profitable for everyone.

When Malaney gave Weinstein this problem, Weinstein took it seriously. Making some simple mathematical assumptions, similar to the ones Coase made, Weinstein soon saw his way to a solution — just the solution, in fact, that Coase had arrived at. But this, Weinstein thought, was a surprise. At least in this case, it seemed as though the mathematics was working in the right sort of way, and indeed, it led to what seemed to be a deeply counterintuitive result that nonetheless bore weight. The process felt surprisingly similar to using mathematics in physics: one makes some simplifying assumptions and then uses mathematics to gain insights into a problem that would have otherwise remained intractable. Most importantly, if someone had told Weinstein about Coase's theorem before he had worked on it himself, he would likely have thought that the solution was politically driven, a thinly veiled case for less government intervention, shrouded in mathematics to give the appearance of rigor. But now he saw that matters were not so simple.

His interest piqued, Weinstein began looking for other cases where mathematics was used to reach counterintuitive results in economics. He uncovered several examples. The Black-Scholes equation was one, since it makes use of fairly sophisticated mathematics to get at the heart of what it means to produce and trade an option. Another was Arrow's theorem, a famous result in social choice theory that essentially proves that if you have a group of people trying to choose between three or more options, there is no voting system that can turn the ranked preferences of all of the individuals in the community into a fair community-wide ranking.

Weinstein realized that his criticisms of economics had been misplaced. Mathematics, he now believed, *could* be used productively to understand economic problems. It was an exhilarating realization, because it meant that someone with some mathematical acumen and a background in physics stood a chance at making progress on problems in economics. Soon, instead of looking for cases where mathematics *had* been put to productive use in economics, Weinstein and Malaney started looking for cases where it *hadn't* been put to use — at least, not yet. Together, they happened on the index number problem. The mathematics underlying the CPI is astoundingly simple, given the profound difficulties associated with assigning a number to something so complicated as the value of money to a consumer. It was a perfect place to start.

Weyl's essential innovation, conceptually speaking, was to find a mathematical theory for comparing otherwise incomparable quantities. In his theory, the incomparable quantities were the lengths of rulers at different locations. His solution was to find a way to bring the rulers to the same location, and then just hold them up next to one another to determine their relationship.

But now think of the index value problem, which, at its core, involves comparing different, apparently incomparable quantities. How can you make sense of the value of money to two different people, especially if they have radically different lifestyles? And how do you compare what might seem like a reasonable market basket in 1950 to what would seem like a reasonable market basket in 1970, or in 2010? These problems seemed insurmountable at first to Weinstein and Malaney. But in the context of the mathematical framework that Weyl and his successors had developed, at least one possible solution emerged. All they needed to do was figure out a way to take any two people — say, a lumberjack in 1950 and a computer programmer in 1995 — and put them in the same circumstances so that they could directly compare their preferences and values. It was a strange thing to propose — after all, the conversation between the lumberjack and the programmer might be a little awkward — but from the point of view of Weyl's mathematics, it was the most natural thing in the world. To solve the index number problem, Weinstein and Malaney argued, you need a gauge theory of economics.

One day, late in 2005, Lee Smolin received an unusual e-mail. It seemed to be about economics, which was unexpected, because Smolin didn't know the first thing about economics. Smolin was a physicist. His work was, and continues to be, in a cutting-edge field known as quantum gravity, which consists of people trying to understand how to put the two revolutionary, immensely successful innovations of early-twentieth-century physics — quantum mechanics, which describes very small objects like electrons, and Einstein's theory of gravitation, which describes *really* big objects, like stars and galaxies — together into a coherent framework. This endeavor had nothing at all to do with economics. Or so Smolin thought.

A few months earlier, Smolin had published an article in the magazine *Physics Today,* a semi-popular publication whose goal was to explain new developments in physics to physicists who weren't necessarily experts in the given area. Smolin's article was an attempt to explain why quantum gravity had not produced a researcher like Albert Einstein, who successfully revolutionized physics by thinking far out of the box. The article was a preview of a book Smolin was just finishing, called *The Trouble with Physics.* In both the article and the book, Smolin argued that physics, or rather, quantum gravity research, faced a sociological problem. A group of physicists working on something called string theory, one approach to solving the basic problem of how to combine gravitational physics with quantum physics, had come to dominate the field. When it came time to hire new faculty members into their physics departments, or to dole out research funding, these string theorists tended to give the resources to other string theorists rather than to people working on alternative approaches to quantum gravity.

It was this *Physics Today* article that had prompted the unexpected e-mail. The man who had written the message was Eric Weinstein, now a hedge fund manager and financial consultant in Manhattan. Weinstein agreed with Smolin's assessment of the physics community, based on his years working as a mathematical physicist at Harvard and then at MIT. But he had a bigger point to make, about how sociology could distort progress in academic research more broadly. As far as Weinstein was concerned, the sociology problem in physics was nothing. Economics was ten times worse.

Smolin wanted to hear more. He invited Weinstein to visit the Perimeter Institute, the research institute in Waterloo, Ontario, where Smolin was based. Perimeter was founded in 1999 by Mike Lazaridis, the entrepreneur and founder of Research in Motion, the company responsible for BlackBerry devices. Perimeter was designed as a place to foster research in fundamental physics. It has a strong reputation for open dialogue and discussion among different approaches on basic questions, in large part because of Lee Smolin's influence on the institute from its earliest days. In some ways, Perimeter is an attempt to answer the sociological problem identified in Smolin's book and articles. It was an ideal place for someone with Weinstein's background and interests to present a new approach to understanding economic theory.

Weinstein visited Perimeter in May 2006. He gave a talk on the way in which gauge-theoretic ideas could be important in a new economic theory, presenting the work he and Malaney had done some years before. And then he left. Smolin and others at the institute found Weinstein's ideas compelling. But they were inclined to be sympathetic. These were not the people who needed to be convinced.

Weinstein and Smolin remained in contact, however. Smolin read up on economics to better understand how gauge theory might help. Around the same time, he began to work with other researchers interested in bringing ideas from physics to bear on economics, including Mike Brown, the first CFO of Microsoft and a former chair of the NASDAQ board, Zoe-Vonna Palmrose, an influential accounting professor working at the SEC, and Stuart Kauffman, who had worked on complex systems theory at the Santa Fe Institute, alongside Doyne Farmer and Norman Packard before they started the Prediction Company.

In September 2008, Weinstein visited Perimeter a second time, for a conference on science in the twenty-first century. The talks focused on ways in which scientific research was changing with new funding sources, with new means of disseminating ideas, such as blogs and online conferences, and with new ideas about where research should and could happen, with places like Perimeter and the Santa Fe Institute becoming centers of study outside of the traditional university.

But the future of science was not at the forefront of Weinstein's mind that September. Just a week after Weinstein's talk at Perimeter, the fourth-largest investment bank in the United States, Lehman Brothers, closed its doors after a century and a half of business. At virtually the same time, AIG, one of the twenty largest publicly traded companies in the world, had its debt downgraded, leading to a liquidity crisis that would have toppled the company had the U.S. government not intervened. In early September, the world economy was already on the ropes. As a hedge fund manager and consultant, Weinstein was tuned in to the surprise and panic in the financial industry, and in economics more generally. As far as Weinstein knew, no one had seen this coming. (Sornette had, but he didn't publicize this prediction widely.)

For Weinstein, the unexpectedly dramatic failure of the U.S. banking system was only further evidence that it was time to take the next step in the development of modern economics. It was time to reflect on what had gone wrong with the now-toxic securities and recognize that economics needed a new set of tools. As physicists had done a generation before, economists needed to broaden their theoretical framework to account for a wider variety of phenomena. Economics needed a new generation of theories and models, suited for the complexity of the modern world. Weinstein thought that the crisis should be an opportunity to set aside past differences between the various approaches to finance and economics. He called for a new large-scale collaboration between economists and researchers from physics and other fields. It would be, he said, an economic Manhattan Project.

Social Security, technically the U.S. federal Old-Age, Survivors, and Disability Insurance program, was first signed into law in 1935 as part of the New Deal, Franklin Roosevelt's program to end the Great Depression through stimulus spending and a broad expansion of the U.S. welfare system. It was a way for the federal government to provide support to the elderly, to children whose parents had died before they were of employable age, and to people who became disabled and unable to work. It was designed to pay for itself, as a real insurance program would. Workers would contribute to the program through a mandatory tax, and the funds collected would be used to pay for the program's costs.

The program was highly controversial. Early on, it was challenged several times in the Supreme Court (unsuccessfully). But over time, as successive generations contributed during the course of their working lives, most Americans came to count on the program as a retirement and disability benefit. By the 1960s, it had become a part of American life, something that workers nearing retirement took as an entitlement. This made matters politically difficult when, during the period of high rates of inflation and low economic growth in the 1970s, it became clear that Social Security was in trouble. Projecting forward, politicians and economists realized that over the coming decades, ever-larger numbers of aging Baby Boomers (then just coming into their own) would retire, and the costs of providing them with benefits would rapidly outstrip the program's ability to fund itself.

And yet, there was little to be done. For a politician to draw attention to Social Security's woes was suicidal. The two obvious solutions to the funding problem — reducing benefits and raising taxes — were equally unappealing. Social Security presented a kind of political catch-22 — that is, until Daniel Patrick Moynihan and Bob Packwood, the two leading members of the Senate Finance Committee in the mid-nineties, shared a moment of inspiration. If you wanted to come up with $1 trillion without anyone noticing the difference, all you needed to do was change the value of money.

Here's how the plan worked. Projections for the future costs of Social Security were based on the expected rate of inflation, which in turn was based on the CPI. Moynihan and Packwood realized that if the official rate of inflation could be lowered, the income from the Social Security tax would rise, and the costs of administering the program would fall. The effect would be to raise taxes and reduce entitlements, relative to the real buying power of money, without acknowledging that you were doing so. The challenge was to find an argument for *why* inflation calculations should be modified. This is where the Boskin Commission came in. It was a masterful sleight of hand. Working backward from the figure of $1 trillion, which Moynihan believed would be necessary to keep Social Security solvent, he and Packwood determined that inflation would need to be reduced by 1.1%.

According to notes written by Robert Gordon, an economist at Northwestern University and one of the five members of the commission, Dale Jorgenson — the Harvard economist who had thrown Malaney out of his office — reported to the commission early on that they were aiming for $1 trillion in Social Security savings over ten years, and that this meant they needed to come up with the requisite reduction in inflation. Then the committee broke up into two teams to work on different ways in which the problems of changing preferences and changing quality could affect CPI. Gordon and the other person on his team, working together, arrived at one number. The other team, which included Jorgenson and Boskin, arrived at another. And then, "somehow" (Gordon's word), when the two teams combined their results, the commission's final recommendation "corrected" inflation by precisely 1.1%.

The Boskin Commission's findings were criticized from all corners. As Gordon later reported, the project was rushed and careless. He and his collaborator finished their contribution days before the commission was due to present to the Senate. The calculations were what physicists and economists both call "back of the envelope," little more than informal estimates. The commission's report was never peer-reviewed before it was presented to the Senate. None of the other members of the commission ever asked how his team had come up with their number, or how the others had arrived at theirs. The answer to such questions would have been inconvenient. (Ultimately, many of the Boskin Commission's recommendations were squashed by effective lobbying on the part of the AARP and others; about five years later, the National Academy of Sciences and the U.S. Bureau of Labor Statistics returned to the problem of how to calculate the CPI, with a more intellectually rigorous approach, and with more nuanced findings.)

Malaney approached Jorgenson with her and Weinstein's ideas about index numbers soon after the Boskin Commission was formed. Jorgenson may have had deep criticisms of Malaney and Weinstein's proposal. They may have even been good criticisms. But it is hard to avoid guessing that it would have caused problems for the Boskin Commission had a new and mathematically rigorous method emerged for calculating precisely what they were tasked to calculate. It seems the easiest thing was to make Malaney and Weinstein go away.

Exporting gauge theories, or other ideas from physics, to economics remains a hard sell. Weinstein was right that late 2008 presented a unique opportunity for someone inclined to change the way economists thought about the world — and the world, economics. Many people in finance, in economics, and in ordinary homes around the world were scared. Things that many people thought they understood turned out to be changing and unreliable. Meanwhile, people working in other fields, such as physics and mathematics, saw an opportunity to contribute to a field that seemed besieged. The suggestion that it was time to reevaluate some of the principal theories and methods of modern economics struck a chord with many, including Smolin and a handful of other physicists working at Perimeter.

Smolin, who previously had been reading up on economics in his spare time, began to consider working on it more seriously. He collected notes he had written on various topics, including his take on Weinstein and Malaney's proposal, and put them together in a paper that he then posted on an online archive where physicists post new research. The paper was a kind of translation dictionary, explaining basic economics to physicists and then showing how ideas that physicists were already comfortable with could be applied to this otherwise foreign science.

Meanwhile, Smolin and Weinstein began working with Smolin's other collaborators—Mike Brown, Zoe-Vonna Palmrose, and Stuart Kauffman—to organize a conference at Perimeter. It was scheduled for May 2009.

The plan was to invite representatives from across the spectrum of economics, to bring together a diverse and heterodox group of people to discuss how to move the field forward after the recent crisis. In addition to the organizers, Doyne Farmer and Emanuel Derman participated. So did some mainstream economists, such as Nouriel Roubini of New York University and Richard Freeman of Harvard, as well as Nassim Taleb. Richard Alexander, a well-known evolutionary biologist, was invited to describe how biology and human behavior could inform economics. The plan was simple. Get a large group of smart people in a room, get them all to see that economics had clear problems, and convince them to work together to come up with a new theory. The plan was to use this conference to kick off the new Manhattan Project.

The conference itself was a success: this wide-ranging group of physicists, biologists, economists, and finance professionals found much to debate and discuss. But when the conference ended, the researchers went their separate ways. As Smolin later explained, there was too much bullheadedness even among these economics outsiders to produce fruitful collaboration. Everyone agreed that economic theory faced major problems, but it was impossible to build consensus on what the problems were, never mind how to fix them. Many of the participants in the conference — as well as other commentators from economics and finance — didn't even agree that a concentrated effort to improve the sophistication of economic modeling was called for. In the background were questions about funding — if the project were funded, how would money be doled out to the participants? — that made the individuals involved cautious of supporting the larger project, for fear they wouldn't receive their cut. And so with regard to the larger goal of creating a new community of interdisciplinary researchers devoted to tackling problems in economics from new directions, the conference failed. After a few months, Smolin gave up on economics and turned his attention back to physics. Now, when he finds himself with a few free minutes, he works on climate science. Economics, he has decided, is intractable — not for the subject matter, but because the field does not seem open to new ways of thinking. Weinstein was right: economics *is* ten times worse than physics.

Today, Weinstein and Malaney continue to work on expanding the mathematical foundations of economic theory. Sornette continues to develop his predictive tools. Farmer is back at the Santa Fe Institute, developing new connections between complexity science and economic modeling. Despite this brainpower, the world economy is in pieces, still bloodied by the 2007–2008 collapse. Can anything be done?

Epilogue: Send Physics, Math, and Money!

I BEGAN THINKING ABOUT this book in the fall of 2008, in the midst of the financial meltdown. At the time, I was about eight months away from a PhD in physics. After a few weeks of researching, I mentioned what I had uncovered to my dissertation advisor. His reaction surprised me. He was convinced, from my examples of how ideas from physics had been used to understand financial markets, that there was a strong connection between the fields. (This, I have found, is the case with most physicists.) But this didn't move him. Instead, he responded by saying that no matter how many physicists had influenced finance, it was impossible to do science on Wall Street.

This idea can be put in different terms. Science isn't a body of knowledge. It's a way of learning about the world — an ongoing process of discovery, testing, and revision. My thesis advisor's reasons for thinking this process couldn't occur on Wall Street were mostly sociological: investment banks and hedge funds are usually very secretive, which means that new ideas developed by such firms are rarely aired and debated the way that new developments in scientific fields are. When a physicist or biologist develops some new insight, he submits a paper on it to a professional journal, where it then undergoes peer review — a process by which new scientific ideas are vetted by other scientists before appearing in print. If a paper passes this first hurdle, it is then scrutinized by the larger community of scientists. Many ideas don't survive this ordeal — they are either never published, or else they languish in obscurity. Even the ideas that are taken up by the community, the ideas that prove most useful, are not accepted as sacrosanct. Instead, they form the starting point for the next generation of theories and models.

In other words, thinking like a physicist is different from (merely) using mathematical models or physical theories. It's how you understand the models that counts. In early 2009, Emanuel Derman, the former physicist who worked with Fischer Black at Goldman Sachs during the eighties and nineties, teamed up with Paul Wilmott, founder of Oxford University's program in quantitative finance, to pen the "Financial Modelers' Manifesto." Their point was in part to defend mathematical models as essential to thinking about finance and economics, and in part to chide "the teachers of finance" who have forgotten that no model states laws by which markets must abide. As they put it, "Models are at bottom tools for approximate thinking." They are never the final word—they rely on assumptions that never hold perfectly, and that sometimes fail entirely. Appropriate use of models requires a good dose of common sense and an awareness of the limitations of whatever model you happen to be using. In this way, they are like any tool. A sledgehammer may be great for laying train rails, but you need to recognize that it won't be very good for hammering in finishing nails on a picture frame.

I believe the history that I have recounted in this book supports the closely related claims that models in finance are best thought of as tools for certain kinds of purposes, and also that these tools make sense only in the context of an iterative process of developing models and then figuring out when, why, and how they fail—so that the next generation of models are robust in ways that the older models were not.

From this perspective, Bachelier represents a first volley, the initial attempt to apply new ideas from statistical physics to an entirely different set of problems. He laid the groundwork for a revolutionary way of thinking about markets. But his work was littered with problems. Most obvious, from the point of view of Samuelson and Osborne, was that the normal distribution he described for stock prices worked only under the very unusual circumstances that prevailed at the Paris Bourse, where there was very little variation in prices. Correcting this problem led to Osborne's hypothesis that returns, not prices, are normally distributed. Mandelbrot's realization that normal and log-normal distributions cannot capture the full wildness of financial markets, then, didn't represent some crisis at the foundations of financial theory, despite his and others' claims to the contrary — rather, it was the first recognition of how Osborne's version of the random walk hypothesis would run aground. That most economists (and physicists interested in such things) now believe that Mandelbrot, too, wasn't quite right is simply another iteration still.

Thorp and Black showed investors how to use the tools developed by Bachelier, Osborne, and Mandelbrot in day-to-day trading — by drawing on still more sophisticated ideas from physics. In some sense, these two scientists are the most important in this book, both because of their pivotal role in putting cutting-edge theory into practice and because they reveal what is involved in using one set of models to build new ones. The options pricing models that Thorp and Black and Scholes developed were based on Osborne's version of the random walk hypothesis, not on Mandelbrot's. This meant that these options pricing models should have been recognized, from the very beginning, as tools with a limited range of applicability. From a physicist's point of view, or an engineer's, starting with Osborne's model made perfect sense. It was far better understood than Mandelbrot's, and so, by adopting a simpler approximation of how market returns really work, Thorp and Black and Scholes were able to turn an extremely difficult problem into a tractable one.

But there was little doubt, even from the beginning, about how these early options pricing models would fail, given Mandelbrot's work: they would misprice extreme events. (Black seemed to recognize the shortcomings of his model as well as anyone — in a 1988 article called "The Holes in Black-Scholes," Black explicitly listed the unrealistic assumptions that went into deriving his formula and described how each of these could lead to errors.) Careful investors, like Michael Greenbaum and Clay Struve at O'Connor and Associates, were able to use their understanding of when the Black-Scholes model would fail to profit and, even more importantly, to protect themselves during the 1987 market crash.

Yet still, the process continues. Both the scientists behind the Prediction Company and Didier Sornette show how new developments in physics can be used to fill in gaps in the random walk, efficient markets thinking behind the Black-Scholes model. The Prediction Company did this by using black box models to identify local, short-term inefficiencies and capitalize on them as quickly as possible — essentially using physics to be the most sophisticated investors in the market. Sornette, meanwhile, has taken Mandelbrot's observation that in wildly random markets, extreme events like market crashes have dominating effects, and asked whether it is possible to predict these catastrophes. The tools he has adapted from seismology go a long way toward showing that dragon kings can be seen from afar.

It is tempting when writing a work of history to try to force the pieces into an overarching narrative. Here, I think, there is a narrative — but it would be a mistake to push it too far. The Prediction Company and Sornette represent two natural and important ways to move forward from the still-dominant Black-Scholes-style thinking. But despite the successes of such models, they are hardly the end of the story. Instead, they are just two examples of particularly fruitful ideas about financial markets — ideas that themselves should be subjected to careful testing and analysis. It's not easy to say what the next major advance will look like: it might be a new way of understanding and anticipating extreme events; it might equally well be a novel test for when the predictions of models are "robust" against inherent market uncertainty; or perhaps it will be a breakthrough in our ability to identify the underlying chaotic patterns lurking in market data. What we *do* know for sure is that there will be a next major advance, and that when we figure out where Sornette's models fail, or where the kind of black box modeling the Prediction Company developed runs aground, we will understand markets more clearly than we do today.

If physicists have been successful at improving our understanding of finance, it is because they have approached problems in a novel way, using methodological insights that are commonplace in physics (and engineering) and that are useful in studying virtually anything. The stories in this book show the methodology in action: one uses simplifying assumptions to make a problem tractable and solve it. Then, once you see how your solution works, you can double back and begin asking what happens when you play with your assumptions. Sometimes you realize that your original solution is no good, because it depends too heavily on assumptions that never really apply; other times, you discover that the solution is pretty good but can be improved in simple ways; and other times still, you realize that your solution is great under certain circumstances, but you need to think about what to do when those circumstances don't apply.

Obviously, physicists aren't the only people who have thought about understanding the world in this way. This kind of model building is ubiquitous in economics and in other sciences. Unsurprisingly, most advances in economics have been made by economists. But physicists are very good — perhaps especially good — at thinking like this. And they are usually trained in a way that helps them solve certain kinds of problems in economics, without the political or intellectual baggage that sometimes hampers economists. Plus, physicists have often come to these problems with different knowledge and backgrounds from people who are trained as economists, which has meant that in some cases, physicists have been able to look at problems in a fresh way.

However, when I say that science is a process, and particularly that financial modeling should be understood as an example of that process, I do not mean to say that financial modelers are somehow marching along the path of scientific progress, inexorably approaching some "final theory" of finance. The goal isn't to find the final theory that will give the right answer in every market setting. It's much more modest. You're trying to find some equations that give you the right answer some of the time, and to understand when they can be relied on.

Derman and Wilmott, in their Manifesto, make this point quite clearly. We should never mistake a good model for the "truth" about financial markets. The most important reason for this is that markets are themselves evolving, in response to changing economic realities, new regulations, and, perhaps most importantly, innovation. For instance, the Black-Scholes model forever changed how options markets operate — which meant that the markets the model was designed to describe were revolutionized by the increasing use of the model. This led to a feedback loop that wasn't fully recognized until after the 1987 crash. As sociologist Donald MacKenzie has observed, financial models are as much the engine behind markets as they are a camera capable of describing them. This means that the markets financial models are trying to capture are a moving target.

Far from undermining the usefulness of models in understanding markets, the fact that markets are constantly evolving only makes the iterative process I have emphasized more important. Suppose that Sornette's model of market crashes is perfect for current markets. Even then, we have to remain ever vigilant. What would happen if investors around the world started using his methods to predict crashes? Would this prevent crashes from occurring? Or would it simply make them bigger, or harder to predict? I don't think anyone knows the answer to this question, which means that it is just the kind of thing we should be studying. The biggest danger facing mathematical modelers is the belief that today's models are the last word on markets.

Weinstein and Malaney's proposal is different from the other ideas discussed in this book. Every other chapter concerns, in one way or another, finance and financial modeling. The other physicists I discussed were looking at a bunch of statistics — stock prices, market moves, annual returns — and trying to make predictions about how the numbers would change in the future. The details of how markets work are of course relevant to such predictions, but it is not so hard to see how, as Osborne observed, a person trained as a physicist is well suited to interpret statistical data. Weinstein and Malaney, however, have proposed a new theory of welfare economics, inspired by ideas developed in physics. This is a far more ambitious project, and one that is more difficult to wrap one's head around.

Nonetheless, if one understands the connection between physics and finance in the right way, there is nothing weird about using physics as a way of making progress in economics more broadly. It isn't that financial markets bear some special connection to the subject matter of physics, or that physics and mathematics can be applied to clearly numerical areas of economics like finance, but not to other areas. Instead, it's that physicists have successfully applied a way of thinking about the world to some areas of economics, and one should fully expect that those methods should be helpful in other areas, too. Indeed, they already *are* helpful in other areas of economics, insofar as economists already use mathematical models for all sorts of things that have nothing to do with finance. Weinstein and Malaney's ideas underline the fact that mathematical tools are used in *every* area of economic thought, including things like policymaking—as shown by the disastrous Boskin Commission.

From this point of view, Weinstein and Malaney's proposal is just a recognition that there are ways to make those models better, to use more powerful mathematics to avoid having to make strong assumptions about people and markets. It may turn out that gauge-theoretic methods are a dead end. But there's no reason to rule them out in advance of careful study. After all, gauge theory was useful in physics when it became clear that a new generation of theories was necessary. One might as well see if it can do the same thing for economics. Weinstein, Malaney, and Smolin have shown that this might be possible.

The idea that methods from physics can be useful in economics is an important one. Equally important is that Weinstein and Malaney's ideas were never given a fair hearing by economists or policymakers. There is something deeply troubling about the suggestion that sociological and financial forces have suppressed an important new discovery that could change how we understand something as crucial to the economy as inflation. With this in mind, Weinstein's Manhattan Project should not be taken as a call for new tools for investors. No one thinks we should devote public resources to the search for a new options model that would help a handful of companies profit. Instead, the proposal was intended to bring mainstream economics up to speed with modern physics and mathematics, setting aside the powerful political and financial forces that distort the discipline.

In a 1965 Supreme Court decision on freedom of speech, Justice William Brennan coined the expression "marketplace of ideas" to describe how the most important insights might be expected to arise out of a free and transparent public discourse. If this is right, then you would expect that the best new ideas about economics would get taken up—even if powerful economists rejected them. This should be particularly true for ideas in finance, since a good idea there can lead to large profits. In this regard, it is interesting that most of the physicists described in the last three chapters of the book, notably Farmer, Packard, and Weinstein, took their ideas to the financial markets when they were rejected by economists. That the ideas have been profitable should be a sign of their importance—and yet, many economists have refused to take notice, including those who set government policy. If there is, as Brennan suggests, a marketplace of ideas, it is deeply inefficient, to the detriment of the rest of us. Smolin moved on to other projects once he realized that mainstream economists weren't interested in hearing what he had to say. Even Sornette, who has worked tirelessly to present his ideas in a way that mainstream economists could understand and appreciate, has hardly been embraced by that community. Much of his audience consists of practitioners.

I don't know how to change the sociology of economics departments. But I think that Weinstein's idea of a major interdisciplinary research initiative would be an excellent start—provided there was some strong institutional or government support behind it, to hold the community together and keep it on track. After all, the original Manhattan Project was a military affair, and it revolutionized physics by changing the way physicists thought about their discipline. A similar commitment on the part of the government or a major not-for-profit in the service of a new generation of economic models would surely have similar effects. More importantly, it would be a source of sorely needed new insights. After years of recession and lackluster growth, it's time to get creative.

When Weinstein first proposed a new Manhattan Project to better understand economics, he was quickly drowned out by the same voices that were criticizing mathematical models and the role of physicists in finance more generally. Indeed, in the years since the 2008 market crash, we've heard a steady drumbeat of criticism of the role of physicists in finance and economics. Words like *quant*, *derivative*, and *model* have taken on some nasty connotations. Now that I have laid out the history of these ideas, the naysayers merit some further thought. It seems to me that if you think about mathematical modeling in the right way, these criticisms are wrong-headed. Understanding why is particularly important because the problems with these criticisms reveal why we should reconsider Weinstein's proposal.

One of the most prominent arguments against mathematical modeling in finance might be thought of as an argument from psychology and human behavior. The idea is that ideas from physics are doomed to fail in finance because they treat markets as though they're composed of things like quarks or pulleys. Physics is fine for billiard balls and inclined planes, even for space travel and nuclear reactors, but as Newton said, it cannot predict the madness of men. This kind of criticism draws heavily on ideas from a field known as behavioral economics, which attempts to understand economics by drawing on psychology and sociology. From this point of view, markets are all about the foibles of human beings — they cannot be reduced to the formulas of physics and mathematics.

There is nothing wrong with behavioral economics — it is clear that a deeper understanding of how individuals interact with one another and with markets is essential to understanding how an economy works. But a criticism of mathematical modeling based on behavioral economics trades on a misunderstanding.

Using physics as a springboard for new ideas in finance does not involve describing people as though they were quarks or pendulums. Think about how the ideas discussed in this book have made the move from physics into financial modeling. Some physicists, like Mandelbrot and Osborne, made progress in understanding markets by simply drawing on their familiarity with statistics to identify new ways of thinking about markets and risk. Others, like Farmer and Packard, used their expertise at extracting information from a noisy source to identify local patterns that could be useful for trading. Still others, like Black, Derman, and Sornette, combined their observations about the details of markets in action with theoretical expertise learned in physics to come up with mathematical expressions that describe how readily observed features of markets (like stock prices and fluctuations) relate to more opaque features (like options prices and oncoming crashes). None of these examples involves assuming that investors are a bunch of quarks or that firms behave like exploding stars.

There's a deeper issue here, however. A careful study of human behavior is hardly inconsistent with using mathematical models to study markets and the economy more broadly. Indeed, psychology, in the form of the Weber-Fechner law, played an important part at the very beginning of mathematical modeling of stock prices: Osborne used it to explain why stock prices exhibited a log-normal distribution and not a normal distribution. More recently, Sornette has shown how accounting for herding effects — another important aspect of human psychology, and a mainstay of the behavioral economics community — can be useful in predicting financial calamity using mathematical techniques. In both of these cases, an understanding of psychology has played a crucial role in developing and refining mathematical models. In general, one should expect studies of psychology and human behavior to be symbiotic with mathematical approaches to economics.

A second kind of criticism — one that has already come up in the book — has found its biggest champion in Nassim Taleb. Taleb has written an influential book, *The Black Swan,* which argues that markets are far too wild to be tamed by physicists. A black swan, you'll recall, is an event that is so unprecedented it is simply impossible to predict. Black swans, Taleb argues, are what really matter — and yet they are precisely what our best mathematical models are unable to anticipate. This is a particular problem for financial modeling, Taleb says. He argues in his book and in many articles that physics lives in a world he calls "Mediocristan," whereas finance lives in "Extremistan." The difference is that randomness in Mediocristan is well behaved and can be described by normal distributions. In Extremistan, normal distributions are simply misleading. For this reason, he argues, applying ideas from physics to finance is a fool's errand.

On one level, what Taleb says is certainly true — and absolutely essential to recognize, especially for people who rely on mathematical models to make real-world decisions. We will never be able to predict everything that can happen. For this reason, a measure of caution and a good helping of common sense are always going to be important when we try to use models successfully. But recognizing that we will never be able to predict everything, and that we shouldn't assume our models reveal some deep truth about what can and cannot occur, is part and parcel of what I have described as thinking like a physicist — it amounts to resisting complacency in model building. And indeed, trying to figure out how to predict the kinds of events that might have seemed like black swans from the perspective of (say) Osborne's random walk model is precisely what led Sornette to start thinking about dragon kings. Surely not every black swan is really a dragon king in disguise. But that shouldn't stop us from figuring out how to predict and understand as many kinds of would-be black swans as possible.

Taleb, though, wants to go further than this. He believes that black swans show that mathematical modeling, in finance and elsewhere, is fundamentally unreliable. Figuring out how to predict dragon kings, or using fat-tailed distributions to address the fact that extreme events occur more often than normal distributions indicate, isn't enough. It seems to me that one can argue successfully that any particular model is flawed — albeit usually in ways that a responsible model builder would recognize from the start. But taking this to the next level and arguing that the model-building enterprise as a whole is doomed is a different matter.

Just consider: the process of building and revising models that I have described here is the basic methodology underlying *all* of science and engineering. It's the best basic tool we have for understanding the world. We use mathematical models cut from the same cloth to build bridges and to design airplane engines, to plan the electric grid and to launch spacecraft. What does it mean to say that the methodology behind these models is flawed — that since it cannot be used to predict everything that could ever happen, it should be abandoned altogether? If Taleb is right about mathematical models, then you should never drive over the George Washington Bridge or the Hoover Dam. After all, at any moment an unprecedented earthquake could occur that the bridge builders' models didn't account for, and the bridge could collapse under the weight of the cars. You should never build a skyscraper because it might be hit by a meteor. Don't fly in an airplane, lest a black swan collide with one of its engines.

Taleb would have it that finance is a different kettle of fish from civil engineering or rocket science, that extreme events are more unpredictable or more dangerous there. But it's hard to see why. Catastrophic events, when they occur, usually come without warning. This is true in all walks of life. And yet, it doesn't follow that we shouldn't do our very best to understand what risks we can, to domesticate as many unknown unknowns as possible. It's important to distinguish between the impossible and the merely very difficult. There's little doubt that mastering financial risk is extremely difficult — much more difficult, as Sornette would say, than solving problems in physics. But the process that I have described in this book is the best way we have ever come up with for addressing our biggest challenges. We shouldn't abandon it here.

There's a third criticism of financial modeling that one sometimes hears. This one is a little deeper. It has been made most influentially by Warren Buffett, who has famously warned of "geeks bearing formulas." This view has it that financial innovation is a dangerous thing because it makes financial markets inherently riskier. The excesses of the 2000s that led to the recent crash were enabled by physicists and mathematicians who didn't understand the real-world consequences of what they were doing, and by profit-hungry banks that let these quants run wild.

There is much that is right in this criticism. The idea that derivatives, including options, are a manufactured "financial product" has proved extremely powerful — and profitable. Over the past forty years, financial engineers have come up with ever more creative, and often convoluted, derivatives, engineered to make money in a wide variety of different circumstances. Dynamic hedging — the idea behind the Black-Scholes model — is the basic tool used in this new kind of banking, since it allows banks to sell such products with apparent impunity. As the banking sector has evolved to put more and more emphasis on new financial products, the impact of a failure of the mathematical models undergirding these products has become ever larger. And indeed, some of these creative new financial products were at the heart of the 2008 crisis. So it is certainly true that physicists and mathematicians enabled a novel kind of risk taking on the part of banks, and that we are now suffering the consequences.

Still, it's not as though market crashes or speculative bubbles are a new phenomenon — after all, the largest market collapse in modern times occurred in 1929, long before derivatives became important. What's more, for the past forty years, essentially the period over which financial innovation has been most important, the financial services sector has buoyed Western economies. In the United States, for instance, the financial services industry has grown six times faster than the economy as a whole. This rapid growth has occurred at the same time that other industries, such as manufacturing, have either declined or grown much more slowly. Financial innovation, like other technological innovation, has thus played a major role in buoying the U.S. and other Western economies over the past three decades. Moreover, there is broad agreement among economists that a large, well-developed financial sector generally spurs growth in other areas of the economy — at least to a point. There is also some evidence that if the financial sector becomes *too* large — as, indeed, perhaps it has — it can negatively impact growth in other areas, largely because finance ends up exerting too much control over other industries. That may well be right, and it may be reason to implement financial reform. But one has to be very careful about throwing the baby out with the bath water: for all sorts of practical reasons, economic growth is a good thing. And worries that the financial sector in the United States or Europe has grown too large hardly undermine the basic point that derivatives, and by extension Black and Scholes's insights, have been essential to producing the growth in the first place. If financial practices had stopped developing in 1975, the world's economies would be far less developed than they are today.

That said, there are many sides to financial innovation. While perhaps some derivatives have spurred growth, many people have criticized their widespread use on account of how complicated they can be, and how difficult to understand. The suggestion seems to be that at least some derivatives are intentionally constructed to confuse or even defraud unsophisticated investors. For instance, this criticism has been leveled against certain derivatives based on consumer loans, such as collateralized debt obligations (CDOs), that played a major role in the 2008 crash. These products involve repackaging mortgages and other loans into derivatives that were supposed to have carefully tailored risks and returns. In part, the reason these particular securities have been so heavily criticized is that many investors, including some major investment banks, were caught off guard when they rapidly declined in value — that is, when they became the "toxic assets" that have plagued U.S. and European banks. There was an enormous amount of confusion about the real risks carried by these products, largely because individual investors were ill equipped to evaluate the risks themselves, while credit ratings agencies like Moody's and Standard & Poor's gave the securities ratings that indicated they were much safer than they turned out to be. To make matters worse, the SEC has charged that Goldman Sachs allowed an outside hedge fund, Paulson & Co., to construct CDOs that were more likely to lose value than their ratings suggested — so that Paulson could bet against the misleadingly risky CDOs.

Surely this episode reveals deep dangers inherent in certain practices involving derivatives. But really, the issue at hand has little to do with derivatives as such. If banks really constructed financial products that looked better than they were just so that their biggest investors could bet against them, as some regulators and others have charged, that is surely unethical. But con artists have been defrauding investors for a long time, without the help of CDOs. It seems to me that derivatives, even CDOs, are best thought of as tools, much like the models used to construct them. Crop futures, for instance, have played an important role in allowing farmers to finance the planting season and to control risk along the way for thousands of years; more recently, currency futures have significantly reduced the risk of international trade, enabling the growth of an international economy. Any tool can be used for more than one purpose — after all, a hammer can be used to hammer a nail, or to bash in a car window. In the hands of police, guns can (at least arguably) be an important part of maintaining a safe and orderly society; yet obviously guns are dangerous in other contexts. Figuring out how to adequately regulate and control derivatives is an important, ongoing policy concern. But it's no different in kind from other regulatory problems.

There is still a worry, however, even after you agree that derivatives and their associated models are tools, to be used judiciously or otherwise. Surely there are some tools — the hydrogen bomb, say, if one can think of that as a tool — that are so dangerous that the world would be a better place if they didn't exist. Perhaps derivatives are, in Buffett's words, "financial weapons of mass destruction" — tools that can be used or misused in such destructive ways that no amount of economic growth is enough to counterbalance the risks. One might even think that the 2008 crisis is evidence of the enormity of the dangers of mathematical modeling in finance. I don't think this is right. To see why, it's worth a careful look at what happened in 2007–2008.

In the film *It's a Wonderful Life*, the main character, George Bailey, runs a savings and loan bank. It's a fairly standard kind of bank: customers deposit their money into an account, in exchange for safety and interest; the bank then turns around and lends the money out, usually as mortgages or business loans. This system works well as long as depositors are by and large happy to leave their money in the bank. On George Bailey's wedding day, however, as he and his new wife drive past his bank, they see a crowd of people clamoring to get in. A rumor has spread that the bank is in trouble and the people of Bedford Falls (Bailey's town) want to withdraw their deposits.

Bailey jumps out of the car, realizing there's been a run on his bank. Inside, he explains to the crowd that their money isn't in the building — it's in their neighbors' houses and their community's stores and businesses. The system fails if everyone tries to withdraw at once, since the bank doesn't keep enough capital on hand to reimburse all of the depositors. In a moment of tragic (but characteristic) selflessness, Bailey realizes that he has a pile of cash on hand — his honeymoon money — and offers to pay some of the depositors out of that, so long as they don't ask too much. He has just enough money that, at the close of the business day, the bank has $1 left and they can shut the doors for the night without going out of business. They have survived the run, but at the expense of Bailey's dreams of traveling the world.

Bank runs were fairly common during the Depression, and even more common during the nineteenth century. They were associated with financial panics, periods in which the economy seemed especially uncertain and no one was sure which banks would survive. A small piece of news that a particular bank was endangered could practically ensure that the bank would fail. Today, bank runs in the United States are a thing of the past, because in 1934 the U.S. government instituted the Federal Deposit Insurance Corporation (FDIC), which insures all consumer bank deposits. Now there's no reason to make a run on a bank, even if you think it's failing: your money is insured by the federal government, no matter what happens.

In the Introduction, I described the quant crisis — the week in August 2007 when all of the major quant funds fell to pieces, for no apparent reason. This was the first hint of a coming catastrophe in the world's financial markets. But what caused the quant crisis? In effect, the quant funds were an early casualty of a much larger-scale bank panic that was setting in that summer, and that would last for more than fifteen months. This panic didn't concern consumer banking, which is protected by the FDIC. Instead, it was a panic that affected a shadow banking system that has developed in the United States over the past three decades. The shadow banking system works like normal banking in principle, but on much larger scales — and with no oversight or regulation. It consists of lending between banks and large corporations (including other banks).

When a firm has cash reserves — say, a few hundred million dollars — it needs a place to deposit them, just as anyone else needs a place to deposit cash. Otherwise, the cash doesn't bear any interest, which amounts to a loss of value. So what firms do is deposit their cash reserves with other firms. This is basically a short-term loan from one bank or firm to another. In exchange, the depositor demands some sort of collateral. One standard choice for collateral would be government bonds, which are essentially risk-free and pay a small amount of interest. But there are only so many government bonds in the world, and many people (and other governments) buy them as long-term investments. And so, as firms' demand for places to deposit their cash has grown, banks have had a strong incentive to come up with other assets that they could use as collateral.

Corporate bonds, which are just like government bonds only issued by a corporation, aren't a very good choice because their value tends to be connected with the corporation's stock prices. No one would want collateral that could be highly volatile, or worse, collateral whose value you could try to "game" by looking at how the stock prices are changing. So firms participating in this shadow banking sector wanted to come up with some new kind of asset that worked like a bond, but whose value didn't depend on something that was easy to get information about. The solution they happened on was consumer debt — mortgages, student loans, credit card debt. Now, consumer debt by itself isn't a great choice, because it's possible to predict from a particular individual's history whether that person is likely to default. So instead of using loans as collateral directly, banks took the consumer loans and "securitized" them. This involved combining a large number of loans into a pool, and then slicing the pool up into pieces and selling the pieces as bonds. These new assets — which included CDOs — were designed to work just like government bonds (though they were much riskier). They bore interest so that when firms deposited money with one another, that money wouldn't lose value.

The quant crisis was the first signal that all was not well in this shadow banking system. The whole system was built on the assumption that U.S. housing markets wouldn't decline. When they did, beginning in 2006, the system began to crumble, and when the decline accelerated in 2007, panic set in. Defaults occurred, mainly among homeowners who were already perceived as high risk, the beneficiaries of so-called subprime mortgages. This sudden high default rate, in turn, made the securities based on subprime mortgages rapidly lose value, as no one was sure whether the promised interest rates would be paid. The quant crisis resulted when a small handful of hedge funds were told they needed to put up more collateral for the loans they used to finance their investment, which in turn meant they needed to sell quickly to raise cash. Most of the quant funds used similar methods, which meant they often had very similar portfolios — so that when one fund started to liquidate, it pushed *all* of their holdings, including the ones that were supposed to act as insurance, lower. This rapid and unexpected loss then forced other funds to sell, too, leading to a vicious cycle where everyone involved lost a lot of money. (This is a perfect example of how Sornette's herding effects can lead to crashes.)

The quant crisis, and its reverberations later in 2007, were just the beginning. The next casualty was the eighty-five-year-old investment bank Bear Stearns, in March 2008. Bear Stearns had been a major player in the shadow banking system, producing many of the securitized loans that served as collateral. When the underlying mortgages started to see ever-higher default rates, Bear Stearns's depositors started to get edgy. Starting in the middle of the month, some of Bear Stearns's biggest customers asked for their money back at the same time. First was Renaissance, James Simons's firm, which wanted its $5 billion. Another $5 billion was pulled out by another hedge fund, D. E. Shaw. Soon it was a classic run on the bank, with all of the customers clamoring for their cash. To stem the bleeding, Bear Stearns was forced to agree to a government-backed takeover by another investment bank, J. P. Morgan.

The crisis was just beginning to pick up steam. The real climax occurred at the end of that summer, when Lehman Brothers, another eminent old investment bank, collapsed. This time, the government didn't step in to negotiate a bailout, which only increased the sense of panic. Within just a few days in September, another struggling investment bank, Merrill Lynch, was eaten up by Bank of America. The insurance firm AIG was on the verge of collapse. No banks were willing to lend money, least of all to other banks, whose fortunes were far from certain. The entire shadow banking system froze up, and the financial market collapsed beneath the pressure. By October, 40% of the value of the U.S. stock market had vanished into thin air.

Surely, the misuse of mathematical models played a role in this crisis. The securitization procedure by which subprime mortgages were turned into new products that behaved like bonds was based on a model developed by a statistician named David X. Li. Li's model had a fundamental flaw: it essentially assumed that default on one mortgage wouldn't change the risk of default on other mortgages. As long as the default rate was low, this was a good assumption — a few isolated defaults didn't have much effect on the housing market. But once the default rate picked up, sometime in 2006, the model stopped making sense. When *lots* of people defaulted on their mortgages, housing prices in the default-heavy neighborhoods dropped — leading to still more defaults. Moreover, the rise in defaults indicated some deeper problems in the economy.

But putting all of the blame for the 2007–2008 crisis on Li's model, or even securitized consumer loans, is a mistake. The crisis was partly a failure of mathematical modeling. But even more, it was a failure of some very sophisticated financial institutions to think like physicists. The model worked well under some conditions, but like any mathematical model, it failed when its assumptions ceased to hold. And it seems that the people who were empowered to make decisions about risk management didn't think through when Li's model would fail them. Everyone was making money, and so they threw caution to the winds. But even this is too easy. The crisis was equally a failure of government policy and regulation, since the shadow banking system that ultimately collapsed ran with essentially no oversight. Either regulators didn't know what was happening, they didn't understand the risks, or they believed that the industry would regulate itself. The crisis resulted from failures on all fronts.

It's worth emphasizing once again that just as O'Connor survived the 1987 crash by being a little more sophisticated in how it used its models than anyone else, Jim Simons's Renaissance Technologies returned 80% in 2008 — again by being smarter than the competition. What's the difference between Renaissance and other hedge funds? It's that Renaissance has figured out a way to do what my dissertation advisor claimed was impossible: do science on Wall Street. This has not involved airing its ideas publicly. Indeed, Renaissance is more secretive than most. But its employees haven't forgotten how to think like physicists, how to question their assumptions and constantly search for the chinks in their models' armor. Much of the company's advantage comes from the quality of the people who work there — they are, by all accounts, simply smarter than most other quants. But equally important is the way the firm is structured: it has a large group of dedicated researchers who are given forty hours a week of unstructured time during which they are encouraged to pursue their own ideas. It is embracing its roots, more than anything else, that has allowed the company to flourish where others have not. Renaissance shows that mathematical sophistication is the remedy, not the disease.

As I finish this book, in early 2012, the world economy still hasn't recovered from the 2008 crisis. If anything, it seems poised for another collapse. And no one expects matters to improve any time soon. The Obama administration has issued predictions that unemployment will hover around 8% through the end of 2012, with sluggish growth in GDP. Both political parties in the United States are simply repeating the same tried-and-failed policy proposals they've been trotting out for a generation. And it's not just the United States. Most of southern Europe is on the verge of default on its sovereign debt — and despite Germany's best efforts, it is hard to see how the euro has a future. Even China and India have shown signs of slowing down. Prospects look dim for the world economy. And the most remarkable thing of all is that no one seems to have any ideas about how to fix it.

There's an ancient Latin proverb that I think applies: "*Extremis malis extrema remedia.*" Desperate times call for desperate measures. What we need now, more than anything else, is a new source of economic ideas. This is why it is time to return to Weinstein's proposal for a new large-scale interdisciplinary research initiative. We have mobilized the U.S. and European scientific communities before, and the result changed the world forever. Given the proven track record of applying ideas from physics in finance that I have described in this book, and the promising directions indicated by Weinstein and Malaney's work, it is time to do the same again. This time, though, the goal wouldn't be a new weapon. It would be a new set of tools for the proper functioning of the world's economies.

Consider that throughout the last decades, and especially during the 2007–2008 crisis, the U.S. government, including its major regulatory groups, has always been a step behind even the least sophisticated banks and investment firms. They're three steps behind the real innovators. When, in the lead-up to the crisis, banks failed to account for the risk associated with securitized loans, there was no one there to point out that the shadow banking system was built on a house of cards. It was only in the aftermath of the crash that new banking regulations made their way through Congress — and even then, the new regulations amounted to rudimentary policy changes designed to protect against yesterday's risks.

This situation should be exactly reversed. We are perfectly happy to devote enormous resources to intelligence initiatives and counterterrorism. But the 2008 market crash did at least as much economic damage as 9/11. We should devote the same resources to staving off economic calamity as we devote to protecting ourselves from other risks. Organizations like the Federal Reserve and the Securities and Exchange Commission, even the World Bank, should be the most sophisticated players in the game — and if these groups are not up to the task, we need some new research organization devoted to interdisciplinary economic research to help guide them. The people charged with running the world's economies should be as good as Renaissance. In fact, they should be better.

Acknowledgments

This book has benefited from many conversations with friends and family over the four years since I began thinking about writing it. The people who stand out in my mind as having made especially substantive contributions — through both their ideas and their encouragement — are Illya Bomash, Bianca Bosker, Peter Byrne, Erik Curiel, David Daniel, Nic Fillion, Sam Fletcher, David Grand, Hans Halvorson, Justin Harvey, Ian Jackson, Leslie Jamison, Kent Johnson, Mary Kate Johnson, Tor Krever, Garrett Lisi, Inna Livitz, Sarah Keller Loveday, Pen Maddy, John Manchak, George Musser, Eoghan O'Donnell, A. J. Packman, Rick Remsen, Chris Search, Kyle Stanford, András Tilcsik, Giovanni Valente, Elliott Wagner, Ken Waters, Thomas Weatherall, Matt Weinstock, Scott Wells, and Amy Wuest. I am grateful to them all. The book owes a special debt to John Conheeney, whose thoughts on the history of derivatives markets gave me a toehold to begin from.

I am also grateful to the people who agreed to be interviewed for the book, and who helped me make contact with the book's subjects and their families. Thank you to Doyne Farmer, Pia Malaney, Sally McClenaghan, Joe Murphy, Holly Osborne, Peter Osborne, Lee Smolin, Didier Sornette, Clay Struve, Ed Thorp, and Eric Weinstein. Pia Malaney, Holly Osborne, Melita Osborne, Peter Osborne, Didier Sornette, Ed Thorp, and Eric Weinstein deserve special thanks for reading early drafts of the chapters to which they contributed and offering useful comments for accuracy.

Some friends and colleagues, in addition to providing valuable insights along the way, also read earlier drafts and offered comments. In every case, their thoughts helped to improve the book in material ways. Thank you, Jeff Barrett, Chris Clearfield, Bennett Holman, John Horgan, Clay Kaminsky, David Malament, Matt Nguyen, and Erin Pearson.

This book would never have been written without the help and encouragement of my agent, Zoë Pagnamenta. She has been a constant source of good advice from the first inkling of an idea through to the final manuscript. I was very lucky to find her — not least because she helped sell the book to the most brilliant (and patient) editor I can imagine, Amanda Cook, who deserves complete credit for anything good in it. She has been a saint through the entire process, and she made the book a joy to write. I am also very grateful to the book's publisher, Bruce Nichols, for his support from the beginning and for ably taking the helm at the very end, when Amanda left Houghton Mifflin Harcourt. And thank you, too, to Ashley Gilliam, for all her help along the way.

My family has been very supportive throughout this process, and I am grateful for it. Thank you to my sister, Katie, and my sisters-in-law, Tara, Lauren, and Carolyn, for many welcome distractions. (And Carolyn, I really am sorry about that trip to Joshua Tree.) Thank you to my father-in-law, Dennis O'Connor, for many conversations and great ideas; and to my mother-in-law, Sylvia O'Connor, for commiserating about the trials of book writing. Thank you to Mo, for so many things, but especially for putting me up (and putting up with me) during Hurricane Irene so that I could finish the Epilogue without threat of power outages. And thank you to my parents, Jim and Maureen Weatherall, for making all of this possible. There was a day, Mom, in July 2010, when you sat on our couch in Irvine and shone with your unending optimism; if not for you that day, I would surely have given up. And Dad, there was an evening in September 2011, at dinner at Saporito soon after you finished reading the first full draft. I have never been happier than I was that evening.

Finally, and above all else, I am grateful to my wife, Cailin. She has read every word of every draft of this book, from the initial proposal to the final bibliography. Which is fitting, because every one of them was written for her.

Notes

1. Introduction: Of Quants and Other Demons

xi *"Simons cuts a professorial figure . . .":* Simons declined to be interviewed for this book. Material on Simons and the history of Renaissance is culled from several sources, including Peltz (2008), Greer (1996), *Seed* magazine (2006), Zuckerman (2005), Lux (2000), and Patterson (2010). Simons is unusually forthcoming (as compared to his usual reticence) about how he became a mathematician, and then how he moved from mathematics and physics into finance, in a public lecture he gave at MIT in 2010 (Simons 2010); he describes his early contributions to mathematical physics and geometry in Zimmerman (2009).

xii *"They called it Medallion . . .":* Ax won the Cole Prize in 1967, and Simons won the Veblen Prize in 1976.

"Over the next decade . . .": The numbers on Medallion's past returns are from Lux (2000) and Zuckerman (2005).

"Compare this to Berkshire Hathaway . . .": These numbers are from the 2010 Berkshire Hathaway annual report (Buffett 2010). The year 2010 is the most recent year for which data are available.

"According to the 2011 Forbes *ranking . . .": Forbes* magazine (2011).

"According to MIT mathematician Isadore Singer . . .": Singer made this remark in the introduction to Simons's public lecture at MIT in 2010 (Simons 2010).

"Hedge funds are supposed to work . . .": For more on the history of hedge funds, including their role in the 2008 crisis, see Mallaby (2010). For background on the workings of financial institutions more generally, see Mishkin and Eakins (2009).

xiii *"They have been around for at least four thousand years . . .":* The details of the early history of derivatives contracts come from Swan (2000). The names used in the text are from real Mesopotamian tablets.

xiv *"But when markets opened . . .":* This history of the 2007 quant crisis, including numbers cited below, comes from Patterson (2010), from news articles from August/September 2007 (Patterson and Raghavan 2007; Lahart 2007; Nocera 2007; Ahrens 2007), as well as from academic work on the topic (Gorton 2010; Khandani and Lo 2011).

xv *"Isaac Newton despaired . . ."*: While it is known that Newton suffered some losses in the South Seas bubble, this quote is sometimes disputed. The attribution seems to originate with Spence (1820, p. 368).

"On average . . .": These numbers come from Sourd (2008).

The Medallion returns are from Willoughby (2008). It is worth pointing out that Renaissance's other principal fund, the Renaissance Institutional Equities Fund, which uses strategies similar to those of the other quant funds and which is designed to have a much higher capitalization than the Medallion Fund, did suffer losses of about 1% in 2007 (Strasburg and Burton 2008).

xvi *". . . promoted by Nassim Taleb . . ."*: See Taleb (2004, 2007a).

xviii *"Even the traditionalists suffered . . ."*: Numbers are from Berkshire Hathaway's 2010 annual report (Buffett 2010).

"Jim Simons's Medallion Fund . . .": The Medallion numbers are from Willoughby (2009).

1. Primordial Seeds

2 *"Or so it would have seemed to Louis Bachelier . . ."*: The story told in this opening section takes some liberties, as certain details of Bachelier's life are not well known. In particular, I am following the French historian of statistics Bernard Bru, who has argued that Bachelier almost certainly worked at the Bourse to support himself during his time at the University of Paris, which began in 1892, and during the years after his PhD when he lived in Paris without regular academic employment (Taqqu 2001). However, as Bru admits, there is no concrete evidence of Bachelier's employment at the Bourse. Whatever else is the case, it is clear that Bachelier had an unusual amount of experience with the French financial system when he wrote his dissertation in 1900. A second liberty concerns the idea that Bachelier would have comforted himself in approaching the Bourse by imagining it as a giant casino. Other details provided here — Bachelier's age, the year he arrived in Paris, his family situation — are all well documented. Biographical details provided here and elsewhere in the chapter come principally from the documents collected in Courtault and Kabanov (2002), as well as Dimand and Ben-El-Mechaiekh (2006), Sullivan and Weithers (1991), Jovanovic (2000), Davis and Etheridge (2006), Mandelbrot (1982), Mandelbrot and Hudson (2004), MacKenzie (2006), and Patterson (2010).

"Inside, it was total bedlam": The Bourse was a variety of open outcry system, and it seems that during the brief periods when the brokers would meet in the building for trades, the scene could become quite disordered. Modern open outcry exchanges are certainly "total bedlam." For more on the history of the Bourse, including various pictures of how it functioned, see Walker (2001) and Lehmann (1991, 1997).

3 *"Laid out in front of him . . ."*: This would have been Bachelier's dissertation

(Bachelier 1900), which is presented in both French and English in Davis and Etheridge (2006).

"Louis Bachelier. It didn't ring any bells": Samuelson told the story of his rediscovery of Bachelier's work in numerous places, including his preface to Davis and Etheridge (2006) and in Samuelson (2000). In this latter reference, Samuelson suggests that he might have heard of Bachelier at least once before Savage's postcard arrived. Note that although the version of the story I tell here, in which Bachelier was forgotten until Savage happened upon his 1914 textbook, is the standard one, there are some who have argued that Bachelier was never really as obscure, even in the English-speaking world, as this standard story suggests. See Jovanovic (2000).

". . . a textbook from 1914 . . .": Savage had found Bachelier (1914).

4 *"That distinction goes to the Italian . . ."*: Much of what is known about Cardano comes from his own autobiography, Cardano (1929 [1576]). Several other biographies have been written, including Morley (1854), Ore (1953), and Siraisi (1997), that seek to put his work (both in mathematics and in medicine) in context. For more on the history of probability generally, see Bernstein (1998), Hacking (1975, 1990), David (1962), Stigler (1986), and Hald (2003).

"Cardano wrote a book . . .": The "book" I have in mind is much of what later became the posthumous *Liber de ludo aleae* (Cardano 1961 [1565]).

5 *". . . a French writer who went by . . ."*: For more on de Méré, Pascal, and Fermat, see Devlin (2008), in addition to the works cited above on the history of probability.

7 *". . . a deep philosophical question at stake"*: For sophisticated but readable overviews of the philosophical difficulties associated with interpreting probability theory, see Hájek (2012), Skyrms (1999), or Hacking (1990).

8 *"This result is known as the law of large numbers"*: For more on the law of large numbers, see Casella and Berger (2002) and Billingsley (1995). See also Bachelier (1937).

9 *"Poincaré was an ideal person to mentor Bachelier"*: For more on Poincaré, see Mahwin (2005) or Galison (2003), as well as references therein.

11 *". . . even he was forced to conclude . . ."*: Poincaré's report on Bachelier's thesis can be found in Courtault and Kabanov (2002), and in translated form in Davis and Etheridge (2006).

12 *". . . according to the Roman poet Titus Lucretius . . ."*: See Lucretius (2008 [60 B.C.], p. 25).

"These experiments were enough . . .": The history of the "atomic theory" and its detractors through the beginning of the twentieth century is fascinating and plays an important role in present debates concerning how mathematical and physical theories can be understood to represent the unobservable world. For instance, see Maddy (1997, 2001, 2007), Chalmers (2009, 2011), and van Fraassen (2009). Although discussing such debates is far from the scope of this book, I should note that the arguments offered here for how one should think of the status of mathematical models in finance are closely connected to more general discussions concerning the status of mathematical or physical theories quite generally.

". . . named after Scottish botanist Robert Brown . . .": Brown's observations were published as Brown (1828).

"The mathematical treatment of Brownian motion . . .": More generally, Brownian motion is an example of a random or "stochastic" process. For an overview of the mathematics of stochastic processes, see Karlin and Taylor (1975, 1981).

". . . it was his 1905 paper that caught Perrin's eye": Einstein published four papers in 1905. One of them was the one I refer to here (Einstein 1905b), but the other three were equally remarkable. In Einstein (1905a), he first suggests that light comes in discrete packets, now called quanta or photons; in Einstein (1905c), he introduces his special theory of relativity; and in Einstein (1905d), he proposes the famous equation $e = mc^2$.

13 *". . . curve known as a normal distribution . . .":* For more on probability distributions, and the normal distribution in particular, see Casella and Berger (2002), Billingsley (1995), and Forbes et al. (2011).

15 *". . . Bachelier was essentially unprecedented in conceiving . . .":* For both sophistication and (ultimately) influence, Bachelier is without peer. But in fact, there were some others who either anticipated Bachelier in some ways (most notably Jules Regnault) or else did similar work within a few years of Bachelier (for instance, Vinzenz Bronzin). For more on these other pioneers in finance, see Poitras (2006) (especially Jovanovic [2006] and Zimmermann and Hafner [2006]) and Girlich (2002).

16 *"The efficient market hypothesis was later rediscovered . . .":* See Fama (1965). The efficient market hypothesis is now a central part of modern economic thought; it is described in detail in any major textbook, such as Mankiw (2012) or Krugman and Wells (2009). For a history of the efficient market hypothesis, see Sewell (2011) and Lim (2006). See also the dozens of recent books and articles attacking the idea that markets are in fact efficient, such as Taleb (2004, 2007a), Fox (2009), Cassidy (2010a, b), Stiglitz (2010), and Krugman (2009).

17 "*. . . called* The Random Character of Stock Market Prices": This is Cootner (1964).

18 *"In Cootner's words . . .":* The quote is from Cootner (1964, p. 3).

". . . mathematical physicist and statistician named E. B. Wilson": Wilson was a polymath who made major contributions to many fields, including statistics, physics, engineering, economics, and public health. In some ways, however, his most lasting contributions were pedagogical; his textbooks on vector analysis (Wilson 1901) and advanced calculus (Wilson 1912) became the standards for a generation of American scientists and engineers. Details about his intellectual biography are to be found in Hunsaker and Lane (1973).

"Gibbs is most famous for . . .": For more on Gibbs and his work, see Hastings (1909), Rukeyser (1988), or Wheeler (1988). His student E. B. Wilson, noted above, also wrote a memoir of his interactions with Gibbs (Wilson 1931).

19 "*. . . called* Foundations of Economic Analysis": This is Samuelson (1947). Samuelson's textbook on economics (Samuelson 1948) further extended his influence on American economic thought.

"*. . . economics matured as a science*": The picture of the history and especially the mathematization of economics presented here is heavily indebted to Morgan (2003).

"*. . . in part because of the work of Irving Fisher . . .*": For more on the life and work of Irving Fisher, see Allen (1993).

20 "*. . . physicist turned economist, Jan Tinbergen*": The claim about the origin of the term *model* comes from Morgan (2003). For a brief biographical sketch of Tinbergen, see Hendry and Morgan (1996); for a more detailed discussion of his work, see Morgan (1990).

"*Unlike in physics . . .*": The relationship between models and theories, and in particular how models in economics differ from theories in physics, is the topic of Derman (2011b).

23 "*He circulated a letter . . .*": The letter is reprinted in Courtault and Kabanov (2002).

"*. . . Lévy read Bachelier's final paper*": The paper would have been Bachelier (1941). The version of the story I tell here comes from Taqqu (2001). It is based on contemporaneous notes on Lévy's copy of Bachelier (1941). Lévy himself, in a much later letter to Benoît Mandelbrot, recalls a slightly different story, in which he encountered a reference to Bachelier in Kolmogorov (1931) in 1931 and immediately returned to Bachelier's work. The existence of the 1941 paper, however, with Lévy's annotations concerning a recent reconciliation, suggest that Lévy misremembered. For more on Lévy, see the biographical note in Mandelbrot (1982).

2. Swimming Upstream

25 "*Maury Osborne's mother . . .*": Almost nothing has been written about M.F.M. Osborne, though his contributions to the early study of market randomness are widely recognized. He is mentioned briefly in Bernstein (1993). The biographical material in this chapter comes from numerous interviews with two of his children, Holly Osborne and Peter Osborne; an interview with one of his principal collaborators, Joe Murphy; and especially from documents provided to me by his family. Included among these documents were two autobiographies that he composed for his family in 1987 (Osborne 1987a, b). Holly, Peter, and their sister, Melita Osborne Carter, were generous enough to read an earlier draft of this chapter and check it for accuracy.

"*So you go and collect that horse manure . . .*": This quote comes from the shorter of two autobiographical documents Osborne dictated before his death (Osborne 1987b).

26 "*. . . to work at the Naval Research Lab (NRL) . . .*": For more on the history of the NRL, both before and after World War II, see Allison (1985) and Gebhard (1979).

"*. . . if you took into account both the lift produced . . .*": This paper, Osborne (1951), didn't appear in print for another six years because, even though Osborne had institutional support for working on whatever he liked, he had difficulty finding journals for some of his most interdisciplinary work. The insect flight paper was ultimately published in the *Journal of Experimental Biology.*

27 "*. . . he worked exclusively on his own projects*": He also served as an internal consultant. Other navy scientists could come by his office and ask questions; Osborne was quick enough and creative enough that he was a resource for the lab even though he did not participate directly in its research initiatives. He also helped during the search operation for the U.S.S. *Thresher,* a nuclear submarine that was lost at sea during a depth test in 1963.

28 "*Nylon*": The story of the development of nylon and Du Pont's participation in the plutonium project comes from Hounshell and Smith (1988), Hounshell (1992), and Ndiaye (2007). Additional details concerning the early reception of nylon are from Handley (2000); for background concerning the Manhattan Project, see Baggott (2009), Rhodes (1995), Jones (1985), and Groves (1962). For more on the dawn and development of "big science," see Galison and Hevly (1992) or Galison (1997).

"*As the* Philadelphia Record *put it . . .*": *Philadelphia Record,* November 10, 1938 (Handley 2000).

31 "*. . . Einstein wrote a letter to Roosevelt . . .*": See Rhodes (1995).

"*. . . Nobel laureate Arthur Compton secretly convened a group . . .*": In addition to the references above on the Manhattan Project, see Compton (1956).

34 "*. . . his parents wouldn't let him attend college so young . . .*": Although this is true, Osborne told the story in a slightly different way: When he graduated from high school, he wanted to go immediately to the University of Virginia, but his parents told him that the college catalog said they would not accept such a young student. The following year, when he went to interview at the university, the interviewer told him that they would have been glad to have him at fifteen. After that, Osborne always cited the college catalog story (apparently manufactured by his parents) as evidence that one should never believe what one reads. This independence of spirit was characteristic of Osborne's intellectual life.

"*Osborne began 'Brownian Motion in the Stock Market' with a thought experiment*": See Osborne (1959, pp. 146–47). It is quite easy to imagine the scene occurring much as he describes.

36 "*The rate of return . . .*": The rate of return is usually just called the returns, or sometimes logarithmic returns, by people who work in finance. But I want to be careful to distinguish it from what you might call absolute returns — that is, the total amount of money earned on an investment — since for many people outside of the profession, it is natural to think of the returns on an investment in terms of the amount earned. It is the logarithmic returns, and not the absolute returns, that Osborne argued were normally distributed.

"*. . . something known as a log-normal distribution*": For background on probability distributions, including log-normal distributions, see Casella and Berger (2002) and Forbes et al. (2011).

38 "*. . . principle known as the Weber-Fechner law*": See Osborne (1959).

39 "*. . . he picked up a book by Einstein,* The Meaning of Relativity *. . .*": This is Einstein (1946).

40 "*So Osborne wrote Einstein a letter . . .*": The original letters are kept at the

Einstein archive at the Hebrew University of Jerusalem. Osborne's family provided me with photocopies (Osborne and Einstein 1946).

41 *"Other researchers, such as the statistician Maurice Kendall . . ."* See Kendall (1953) in particular. Kendall's work on the randomness of stock prices is described in detail in Bernstein (1993).

42 *"As Osborne would later put it . . .":* The quote is from Osborne (1987a, p. 137).

"The third idea concerned the migratory efficiency of salmon": This work was ultimately published as Osborne (1961).

44 *"Osborne proposed a new model for deep ocean currents":* This work was published as Osborne (1973).

". . . it was impossible to predict how individual stock prices would change . . .": Osborne makes this point in several places, but he dwells on it (and the related question of how analyses such as his might be put into use in practice) in his book, Osborne (1977, pp. 96–100).

45 *". . . 'unrelieved bedlam' . . .":* See, for instance, Osborne (1962, p. 378) for the quote. For a clear example of where Osborne relentlessly sought empirical evidence *against* his own hypothesis, see Osborne (1967).

"He showed that the volume of trading . . .": See Osborne (1962). Note that this work appeared just one year after the migratory salmon paper was published.

". . . Osborne and a collaborator . . .": The article I have in mind is Niederhoffer and Osborne (1966); the collaborator was Victor Niederhoffer, the now-(in)famous hedge fund manager. For more on Niederhoffer, see his autobiography, Niederhoffer (1998), or the recent *New Yorker* profile (Cassidy 2007).

48 *". . . Osborne proposed the first trading program . . .":* In other words, the first systematic, fully deterministic trading strategy that could be programmed into a computer — a system for what today would be called algorithmic trading. The proposal is made in Niederhoffer and Osborne (1966).

3. From Coastlines to Cotton Prices

49 *"Szolem Mandelbrojt was the very model . . .":* Information about Mandelbrojt comes from O'Connor and Robertson (2005), as well as from the biographical materials related to Mandelbrot cited below.

"In 1950, Benoît Mandelbrot . . .": Unfortunately, Mandelbrot passed away in 2010, before I had an opportunity to interview him in connection with this book. Biographical material in this chapter comes from Mandelbrot and Hudson (2004), Mandelbrot (1987, 2004a), Gleick (1987), Barcellos (1985), and Davis (1984), as well as from a number of filmed interviews of Mandelbrot produced shortly before he died — especially Mandelbrot (1998, 2010).

50 *"This is for you . . .":* This story, including the quote, is told in Mandelbrot and Hudson (2004).

". . . linguist named George Kingsley Zipf . . .": For more on Zipf, see Mandelbrot's biographical notes at the end of Mandelbrot (1982). For the most up-to-date

take on the mathematics of Zipf's law, see Saichev et al. (2010) — a book coauthored by Didier Sornette, who is the subject of Chapter 7 of this book.

51 "*. . . which he named 'fractal geometry' . . .*": For more on fractal geometry, see, for instance, Falconer (2003).

52 "*. . . indeed, in a speech he gave . . .*": This is Mandelbrot (2004a).

53 "*. . . of the more than 3 million Jews who lived in Poland . . .*": Background material on World War II and the Holocaust in particular is from Dwork and van Pelt (2002), Fischel (1998), Rossel (1992), and Yahil (1987).

54 "*How long is Britain's coastline?*": This question is taken up in Mandelbrot (1967).

55 "*. . . a coastline doesn't have a length . . .*": The more precise version of this claim is that a coastline should be understood to have non-integer Hausdorff dimension, which means that the correct "measure" of a coastline does not behave like a length.

"*It was one of his first attempts . . .*": Mandelbrot coined the term *fractal* in Mandelbrot (1975), which was translated into English as Mandelbrot (1977). But Mandelbrot (1967) is one of the first places where he describes geometrical objects with non-integer Hausdorff dimension exhibiting self-similarity.

56 "*. . . but anti-Semitism in the south was less virulent . . .*": While the comparative claim is true, it should not be taken to mean that anti-Semitism was not rampant in Vichy France. For more on Vichy France during World War II, including French anti-Semitism during the war, see Paxton (1972), Marrus and Paxton (1995), and Poznanski (2001).

57 "*. . . except to say that . . .*": These quotes come from the interview that Mandelbrot did for Web of Stories (Mandelbrot 1998).

58 "*In Thomas Pynchon's novel* Gravity's Rainbow *. . .*": This is Pynchon (1973).

59 "*The normal distribution shows up . . .*": Indeed, an important result of mathematical statistics, the central limit theorem, states that if you can model a random variable as the sum of a sufficiently large number of independent and identically distributed random variables, where the distribution of the random variables in the sum has finite mean (average) and variance (volatility), then the random variable must be normally distributed, even if the variables in the sum are not normally distributed. This means that normal distributions appear all over the place. As we shall see, however, Mandelbrot argued that for financial markets, one of the assumptions of the central limit theorem fails: he argues that the distributions of market returns do not have finite variance. For more on the central limit theorem, see Billingsley (1995), Casella and Berger (2002), and Forbes et al. (2011). For more on Mandelbrot's claims, see Mandelbrot (1997) and Mandelbrot and Hudson (2004).

"*. . . the law of large numbers for probability distributions . . .*": It is actually more general than the other version of the law of large numbers, which governs how probabilities for simple games like coin flips relate to frequency. The law of large numbers for probability distributions can be used to prove the other version, as can be seen by thinking about the coin-tossing example.

"Not all probability distributions satisfy the law of large numbers . . .": The more precise version of this claim is that not all distributions have finite mean—and indeed, Cauchy distributions do not have finite mean. For more on Cauchy distributions and the law of large numbers, see Casella and Berger (2002), Billingsley (1995), and Forbes et al. (2011).

61 *"But then 'a storm' would come through . . .":* Mandelbrot describes this aspect of his wartime experience in Mandelbrot (1998).

62 *"This is a general property of fractals . . .":* There are many connections between fractals and fat-tailed distributions. That certain features of fractals exhibit fat tails is one such connection; another is that (some) fat-tailed distributions themselves exhibit self-similarity, in the form of power-law scaling in their tails. Mandelbrot was a central figure in identifying and exploring these relationships. See Mandelbrot (1997).

63 *"Known as the Butcher of Lyon . . .":* For more on Barbie, see Bower (1984) and McKale (2012).

65 *". . . 'there was no great distinction . . .'":* This quote is from Mandelbrot (1998).

66 *". . . and economist named Vilfredo Pareto":* The definitive collection on Pareto and his influence is the three-volume Wood and McClure (1999); see also Cirillo (1979).

68 *". . . it appeared that there was no 'average' rate of return":* In other words, it seemed that neither mean nor variance was defined for the distributions of cotton prices. As described below, Mandelbrot would later argue that the distributions of rates of return for financial markets *do* have finite means, but not variances. However, it can often be difficult to calculate the mean for a Lévy-stable distribution—in cases where variance is undefined, the average value calculated from any finite data set takes a long time to converge to the mean—which accounts for why Mandelbrot and Houthakker originally believed that the mean did not exist.

". . . discovered by one of his professors in Paris, Paul Lévy": Mandelbrot offers some biographical background on Lévy in Mandelbrot (1982) and describes his interactions with him in Mandelbrot and Hudson (2004).

69 *". . . a class of probability distribution now called Lévy-stable distributions":* They are also called α-stable distributions. Throughout the text (and in Mandelbrot's popular writing), "wildness" is code for "$\alpha < 2$." For a Lévy-stable distribution with $1 < \alpha < 2$, the mean is defined, but the variance is not; if $\alpha \leq 1$, neither mean nor variance is defined. Notably, the central limit theorem fails for Lévy-stable distributions, or rather, the following more general theorem holds: a random variable that can be modeled as a sum of sufficiently many independent and identically Lévy-stable-distributed variables must also be Lévy-stable distributed. For more on the mathematics of Lévy-stable distributions, see Mantegna and Stanley (2000) and Zolotarev (1986).

70 *". . . early enough that Paul Cootner . . .":* See Mandelbrot (1964) and Cootner (1964).

71 *"Cootner made the argument . . .":* The passage is quoted in Mandelbrot and Hudson (2004, p. xxiii).

". . . most notably, Nassim Taleb . . .": See Taleb (2004, 2007a). Mandelbrot

makes related arguments in Mandelbrot and Hudson (2004). For a more moderate version of Taleb's argument—one that is sympathetic with the central arguments of this book, though perhaps not regarding the present point on the direction history took in 1962—see Taleb (2007b).

72 *"On a typical day, there aren't going to be any extreme events . . ."*: While true, this remark obscures some important points, often emphasized by Mandelbrot. For one, the statistical tools that one uses in the context of normal and log-normal distributions often do not make any sense—and certainly do not work—in the context of Lévy-stable-distributed variables. For this reason, assuming normal or log-normal distributions can lead to extremely misleading results and, moreover, produce a false sense of confidence regarding the likelihood of certain kinds of extreme events. For another, despite the fact that extreme events happen infrequently on both models, in Mandelbrot's models of financial markets, they happen often enough that it is the extreme events that dominate market behavior in the long run. And so, even if there are similarities in how the models predict markets on a typical day, there is a significant difference in how one should view the importance of a "typical day" for the long-term behavior of markets.

73 *"It is also too simple to say that Mandelbrot was ignored . . ."*: For instance, see Fama (1964).

"Today, the best evidence indicates . . .": See, for instance, Cont (2001) and references therein; this point was also emphasized in conversation by Didier Sornette, whose work is the subject of Chapter 7.

74 *". . . there is disagreement about how to interpret the data"*: In particular, it can be extremely difficult to tell whether empirical data are governed by distributions that are Lévy-stable and distributions that are fat-tailed but not Lévy-stable, since the differences often turn on the frequency of extreme events that occur very infrequently. See, for instance, Weron (2001).

4. Beating the Dealer

76 *"The year is 1961"*: I have taken some liberties with this opening story (Des Moines; whiskey sours), but the basics are correct; it is based on an autobiographical essay (Thorp 1998). More generally, the biographical material on Thorp is from that essay, as well as Thorp (1966, 2004), Poundstone (2005), Patterson (2010), and Schwager (2012). In addition, I interviewed Thorp, and he was kind enough to read and comment on an earlier draft of this chapter.

78 *". . . the 1973 book* A Random Walk Down Wall Street *. . ."*: This is Malkiel (1973).

80 *". . . about $850 in 2012 dollars . . ."*: This calculation is based on the Bureau of Labor Statistics' online inflation calculator at http://www.bls.gov/data/inflation_calculator.htm.

81 *". . . the pioneering mathematician profiled by Sylvia Nasar . . ."*: See Nasar (1998).

82 For more on Shannon, see Kahn (1967), Poundstone (2005), Gleick (2011),

and the two biographies in Wyner and Sloane (1993). An excellent modern introduction to information theory is Gray (2011); for Shannon's contributions specifically, see Wyner and Sloane (1993) and Shannon and Weaver (1949).

85 "*. . . as Shannon's secretary would later inform Thorp . . .*": This quote is from Thorp (1998).

86 "*. . . a colleague passed along a recent academic article . . .*": The article was Baldwin et al. (1956).

"*Cervantes . . . wrote stories in which his characters became proficient . . .*": See "Rinconete and Cortadillo" in Cervantes (1881).

88 "*. . . and then offered to submit Thorp's paper . . .*": The paper was accepted and published as Thorp (1961).

89 "*His name was Manny Kimmel*": Kimmel's life, including the story of how his parking lot business was transformed into the Time Warner empire, is described in Poundstone (2005) and, especially, Bruck (1994). The back story told here is based on those sources; the story of Kimmel's trip to Vegas with Thorp is based on Thorp (1966).

91 "*. . . in a book,* Beat the Dealer *. . .*": See Thorp (1966).

92 "*. . . a paper written by one of Shannon's colleagues . . .*": The paper was Kelly (1956). For more on the Kelly criterion, see Thorp (2006), MacLean et al. (2011), and Thorp (1984, Pt. 4).

"*Kelly was a pistol-loving, chain-smoking, party-going wild man . . .*": This account of Kelly is based on Poundstone (2005), though a review in *American Scientist* disputes this account.

93 "*Imagine you're in Las Vegas, betting on the Belmont Stakes . . .*": For the purposes of illustration, I am intentionally ignoring federal laws on wire-based gambling that were in effect in the 1960s.

94 "*. . . you're guaranteed to win one of them . . .*": Suppose you have $100 to start with, and you put $17 on Epitaph and $83 on Valentine. If Valentine wins, you get back your original $83, plus 5/9 more, for a total of $129. But you also paid $17 for your (lost) bet on Epitaph. So your total profit is $12. Meanwhile, if Epitaph wins, you get your $17 plus 7 times more, for a total of $136, less the $83 you bet on Valentine (and lost). So in this case your profit is $53. In either case you win.

96 "*This would initialize the device . . .*": These details about the computer are based on Thorp (1998).

"*. . . the calculations for that level of precision were far too complicated*": In fact, the calculations for where a ball will fall given just the standard roulette setup — a ball rolling around a rotating wheel — were not too complicated for the computer to solve. However, roulette wheels are designed with small bumps on the wheel that act as randomizers, so that if the ball hits one of these bumps, it will bounce around and change its trajectory. The computer was not able to predict precisely how these randomizers would affect where the ball would land, thus introducing additional uncertainty.

98 "*. . . essays that featured papers by Bachelier, Osborne, and Mandelbrot*": Importantly for the central argument of this book, Thorp confirmed in an interview that he

read the paper by Mandelbrot included in the Cootner volume, in addition to the papers by Bachelier and Osborne. Though he saw how fat tails could affect a model based on the log-normal distribution, he nonetheless decided to use Osborne's simpler model for returns in constructing his options pricing formula. He took tail effects into account by behaving cautiously in the applications of his pricing formula, that is, by remaining cognizant of the fact that it would fail under certain circumstances.

102 *"In 1967, he wrote a book . . ."*: The book was Thorp and Kassouf (1967).

103 *"And Princeton-Newport's demise was particularly dramatic"*: This story, that Princeton-Newport was moving positions off its books to avoid tax losses, is based on the accounts in Poundstone (2005) and Stewart (1992), and in contemporary news stories such as Eichenwald (1989a, b). In an interview, Thorp emphasized another aspect of the allegations, which amounted to stock parking in the other direction: a trader at Milken's firm, Bruce Newberg, was using Princeton-Newport to move positions off of his books to avoid federal reporting laws and Drexel's trading rules. Regan, Newberg, and the other defendants were initially found guilty of both varieties of charges, though the convictions were overturned on appeal and the defendants were cleared of wrongdoing.

104 *"As the* Wall Street Journal *put it in 1974 . . ."*: The article was Laing (1974).

5. Physics Hits the Street

105 *"In February 1961, Fischer Black's PhD advisor . . ."*: This quote is from Mehrling (2005, p. 37). The biographical material on Fischer Black comes mostly from the recent biography, Mehrling (2005), with some material from Black (1987, 1989), Merton and Scholes (1995), Lehmann (2005), Derman (2004, 2011a), Figlewski (1995), Forfar (2007), Bernstein (2010), and Bernstein (1993), as well as from an interview with Emanuel Derman, who worked and collaborated with Black at Goldman Sachs.

"Within a week, Black was in jail for participating in student riots . . .": Strangely enough, the impetus for the riots was Harvard president Nathan Pusey's decision to print diplomas in English, rather than Latin. On one day of rioting, four thousand students demonstrated; Harvard police dispersed them using tear gas and smoke bombs. The sixties had begun.

". . . remains the standard . . .": Whether this position is just is an important question, but that the Black-Scholes model holds a privileged position in the first place seems clear. See Haug and Taleb (2011).

106 *". . . the American Financial Association awards the Fischer Black Prize . . ."*: The quote is from the AFA's website's description of the Fischer Black Prize. See http://www.afajof.org/association/fischerblack.asp.

110 *". . . now known as the Capital Asset Pricing Model (CAPM)"*: Treynor (1961) was not the only person to come up with the CAPM, though it is now widely recognized that he was the first. Others with claims to have developed the CAPM include William Sharpe (1964), who won the Nobel Prize for his contribution to asset pricing in 1990; and John Lintner (1965). See, for instance, French (2003) for more on the provenance of

the CAPM; see also Bernstein (1993).

111 "*'Equilibrium was the concept that attracted me . . .'* ": This quote is from Black (1987, p. xxi).

"*These states are called equilibrium states . . .*" That the concepts of equilibrium in physics and economics are so similar traces back to Samuelson's Gibbsian heritage.

"*. . . an 'interesting fellow,' in Jensen's estimation*": The quote is from Merton and Scholes (1995, p. 121).

113 "*Black's strategy of building a risk-free asset . . .*": It seems that there are several things that go by the name "dynamic hedging," and indeed, any strategy that involves regularly changing one's hedge deserves the name. Throughout the text, however, I will mean something very specific: a strategy by which one constantly updates the proportions of stocks and options in one's portfolio so that the portfolio as a whole is risk-free.

114 "*. . . successfully urged the* Journal of Political Economy *to reconsider . . .*": The article was published as Black and Scholes (1973). See also Merton (1973) and Black and Scholes (1972, 1974). For more on the Black-Scholes formula and its generalizations and extensions, see Hull (2011) and Cox and Rubinstein (1985).

115 "*The head of that committee was James Lorie . . .*": For more on the history of the CBOE, see Markham (2002) and MacKenzie (2006).

"*On the first day of trading . . .*": These numbers are from Markham (2002, vol. 3, p. 52).

"*But volume grew at an astonishing rate . . .*": These numbers are from Ansbacher (2000, p. xii).

116 "*In January 1977, the European Options Exchange was established . . .*": For more on options markets in Europe, see Michie (1999).

"*. . . Friedman wrote him a letter . . .*": This is from Milton Friedman's foreword to Melamed (1993).

"*Bretton Woods, named for the town in New Hampshire . . .*": For more on the Bretton Woods system, see Markham (2002) and MacKenzie (2006), as well as Eichengreen (2008) and Melamed (1993).

117 "*. . . Leo Melamed, the chairman of the Chicago Mercantile Exchange . . .*": For more on the history of the CME and the IMM, see Melamed (1993).

118 "*What does the IMM have to do with Black and Scholes . . .*": I am grateful to John Conheeney, former chief executive of Merrill Lynch Futures and former board member of both the Chicago Board of Trade and the Chicago Mercantile Exchange, for pointing out the relationship between the decay of Bretton Woods and the rise of derivatives trading.

119 "*The distinction may seem inconsequential . . .*": I am grateful to Emanuel Derman for pointing out to me how consequential the differences are, from the perspective of practicing bankers. See, however, Derman and Taleb (2005) and Haug and Taleb (2011).

120 "*This led him to a new theory of macroeconomics . . .*": General equilibrium has its roots in Samuelson (1947), and in his Gibbsian heritage. Black's contributions to the

idea were original, however. See Black (1987) for a collection of essays on this topic, and Black (2010) for his later views on the subject.

121 *"But after* Sputnik *was launched . . ."*: For more on the effects of *Sputnik* on U.S. science, see Wang (2008), Cadbury (2006), and Collins (1999). The data presented here on physics PhDs are from the American Institute of Physics Statistical Research Center, at http://www.aip.org/statistics/. The data on the NASA budget over time are from the Office of Management and Budget, as reported by Rogers (2010).

122 *"Emanuel Derman was a South African physicist who experienced . . ."*: Material on Derman is from Derman (2004, 2011b) and from an interview I conducted with him.

123 *"Beginning with the Carter administration . . ."*: For more on Volcker's war on inflation, see Markham (2002).

"Sherman McCoy . . . was an eighties-era bond trader . . .": See Wolf (1987).

124 *". . . the Black-Scholes model won't get options prices right"*: To his credit, Black understood quite clearly that his model had shortcomings, and that it was at best a first approximation. See, for instance, Black (1992).

125 *"Blame for the crash fell to a novel financial product . . ."*: For more on portfolio insurance, see (for instance) Bernstein (1993). See also Markham (2002).

"Markets themselves seemed to change in the wake of the crash": See MacKenzie (2006).

126 *"The smile appeared suddenly and presented a major mystery . . ."*: Notably, Clay Struve, whom I discuss below, indicated that he and his coworkers were aware of the volatility smile even before the crash of 1987 — that is, it didn't appear so suddenly after all, if you knew to look for it!

". . . Emanuel Derman came up with a way of modifying the Black-Scholes model . . .": See Derman and Kani (1994).

"There's an interesting, and rarely told, twist to the story . . .": This story is based on an interview I performed with Clay Struve, as well as a published interview with Michael Greenbaum (Jung 2007), and Cone (1999). Greenbaum mentions that O'Connor was using jump diffusion models in the late 1970s; Struve confirmed it. Cone (1999), meanwhile, described how Struve saved O'Connor in October 1987.

127 *"Models have failed in other market disasters as well . . ."*: For more on Long-Term Capital Management, see Lowenstein (2000).

6. The Prediction Company

130 *"When the Santa Fe Trail . . ."*: For more on the Santa Fe Trail, see Duffus (1972).

"A century and a half later, two men . . .": The narrative history of the founding of the Prediction Company is from Bass (1999). Additional biographical details concerning the founders of the Prediction Company come from Bass (1985, 1999), Gleick (1987), Kelly (1994a, b), and Kaplan (2002), as well as interviews and e-mail exchanges

with Doyne Farmer and others knowledgeable about the early history of the company.

131 *"In Packard's words . . ."*: The expression "the edge of chaos" comes from Packard (1988).

132 *"As head of the Manhattan Project . . ."*: For more on Frank Oppenheimer, see Cole (2009). For more on J. Robert Oppenheimer, see Bird and Sherwin (2005), Conant (2005), and Pais (2006), as well as the references given above on the Manhattan Project: Baggott (2009), Rhodes (1995), Jones (1985), and Groves (1962).

". . . the Washington Times-Herald *reported . . ."*: This was in the July 12, 1947, issue of the newspaper.

133 *". . . a young graduate student named Tom Ingerson"*: The story of Ingerson, Oppenheimer, Farmer, and Packard is from Bass (1985, 1999).

"Jobs at the top universities were filled . . .": For an example of the kind of recommendation I have in mind, see Wheeler (2011). This letter is the origin of the quote "best men" in the next sentence.

134 *". . . Silver City was a paradigm Western mining town"*: This background on Silver City is from Wallis (2007).

136 *". . . first developed by a man named Edward Lorenz"*: The biographical and historical details concerning Lorenz and the history of chaos theory are from Gleick (1987) and Lorenz (1993).

138 *". . . the work of two physicists named James Yorke and Tien-Yien Li . . ."*: The article is Li and Yorke (1975).

139 *". . . the so-called butterfly effect . . ."*: The paper is Lorenz (2000). Lorenz never used the metaphor of a butterfly flapping its wings, though he sometimes used a similar metaphor involving a seagull.

". . . Farmer through reading A. H. Morehead . . .": Farmer read Morehead (1967); Packard read Thorp (1966).

143 *". . . where the ball lands is sensitive to the initial conditions . . ."*: Although there is some controversy concerning just what should count as a truly chaotic system, virtually everyone would agree that roulette is *not* chaotic. The reason is that the ball and wheel always come to rest in a small number of possible configurations, and so there is a strong sense in which all initial conditions lead to a small number of possible final states. But there is a precise mathematical sense in which roulette is "almost" chaotic, since if you ignore loss of energy from things like friction, the system becomes chaotic. For more on what it means for a system to be chaotic, see, for instance, Strogatz (1994) or Guggenheimer and Holmes (1983).

". . . the Dynamical Systems Collective and the Chaos Cabal": In fact, they published papers with the Dynamical Systems Collective as their "official" affiliation: for instance, see Packard et al. (1980).

144 *". . . these attractors have a highly intricate fractal structure"*: In addition to Strogatz (1994) and Guggenheimer and Holmes (1983), see Mandelbrot (2004b).

"The collective's most important paper . . .": This is "Geometry from a Time Series" (Packard et al. 1980).

146 *"The Santa Fe Institute hosted its first conference on economics . . ."*: The proceedings of these conferences were published as Anderson et al. (1988), Arthur et al. (1997), and Blume (2006).

148 *"Things got even better after the* New York Times *. . ."*: The article was Broad (1992).

149 *". . . this is how their enterprise is usually characterized . . ."*: One certainly gets this impression from Bass (1999); likewise, Broad (1992) writes that Farmer and Packard are "private entrepreneurs using world-class skills in chaos theory to predict the rise and fall of stocks and bonds."

"Farmer and Packard didn't use chaos theory . . .": This section in particular is based on an interview with Farmer. The closest thing from Farmer's and Packard's days as physicists that was helpful in their early days with the Prediction Company was the work in Farmer and Sidorowich (1987), where they present a method for making short-term predictions based on a particular algorithmic approximation.

"One strategy they used was something called statistical arbitrage . . .": For more on the history of statistical arbitrage, see Bookstaber (2007). Ed Thorp also played a significant role in the early development of the idea; for more on his contribution, see Thorp (2004).

150 *". . . a variety of computer programs known as genetic algorithms"*: For more on genetic algorithms, see, for instance, Mitchell (1998). For Packard's early contributions, see Packard (1988, 1990).

154 *". . . over the firm's first fifteen years . . ."*: More specifically, this person told me that the company had a Sharpe ratio of 3.

7. Tyranny of the Dragon King

159 *"Didier Sornette looked at the data again"*: The opening story, which plays out throughout the chapter, is a dramatization, but the basic details are correct. In late summer 1997, Sornette observed a pattern in U.S. financial data that he had previously argued could be used to predict financial crashes; he contacted his colleagues Olivier Ledoit and Anders Johansen and proceeded as described here. The story is told briefly, for instance, in Chapman (1998) and alluded to in Sornette (2003); further details are from an interview and numerous e-mail exchanges with Sornette. More generally, biographical material on Sornette is based on this interview and on Sornette's story of how he became interested in finance in Sornette (2003). Sornette generously read an earlier draft of this chapter and offered helpful comments to improve its clarity and accuracy.

161 *"Imagine inflating a balloon"*: Sornette offers a very clear account of how critical phenomena may be used to understand market crashes, including an explanation of the mechanisms that lead to self-organization in markets, in Sornette (2003). For more on Sornette's work on critical ruptures, and the application of these ideas to other contexts, see also Sornette (2000).

162 *"He has also written four books . . ."*: These are Sornette (2003), Sornette (2000), Malevergne and Sornette (2006), and Saichev et al. (2010).

164 *"But by the mid-1960s, the leaders of several western European nations . . ."*: The definitive history of the European Space Agency, from 1973 through 1987, is to be found in Krige et al. (2000).

166 *"This kind of conspiracy is sometimes called self-organization . . ."*: Self-organization is an old idea, though its modern form originates in the work of the 1977 Nobel laureate for chemistry, Ilya Prigogine (Glansdorff and Prigogine 1971; Prigogine and Nicolis 1977). The idea described in the text is more specifically described as "self-organized criticality," which is due to Bak et al. (1987); see also Bak (1996).

167 *"Together, they began to think about . . ."*: Sauron's thesis, Sauron (1990), involved constructing a small (physical) model of the Earth's crust using sand, Silly Putty, and honey, and then using it to perform experiments on how the crust buckles when plates collide. Sauron and Sornette showed that these collisions exhibited a characteristic fractal pattern. This work is described in Davy et al. (1990), Sornette and Sornette (1990), Sornette et al. (1990a, b), and Sornette (2000). For more on the history of the study of tectonic plates, see Oreskes and Le Grand (2003).

168 *"The ancient Roman historian Aelian . . ."*: Aelian makes these remarks in *On Animals*, translated as Aelian (1959 [200 A.D.]).

"An ancient Indian astrologer . . .": See Bhat (1981).

". . . he and Sauron published a paper . . .": This paper was Sornette and Sornette (1996).

169 *"It was not the very first paper . . ."*: The most important precursors to the Sornettes' idea were Vere-Jones (1977), Allegre et al. (1982), Smalley and Turcotte (1985), and Voight (1988).

"The moment of inspiration came two years later . . .": This moment occurred while Sornette was working for Aérospatiale. It was developed in conjunction with pressure tanks in a series of publications over the following years, beginning with Sornette and Vanneste (1992), followed by Sornette et al. (1992), Vanneste and Sornette (1992), and Sornette and Vanneste (1994). The discovery of the log-periodic acoustic emissions before a critical rupture was first presented in detail in Anifrani (1995). This idea was then tested experimentally, and results were presented in Lamaignère et al. (1996, 1997) and Johansen and Sornette (2000).

170 *". . . predict a critical earthquake . . ."*: Sornette and collaborators first introduced the idea of a critical earthquake in Sornette and Sammis (1995), building on ideas presented by Bufe and Varnes (1993). The idea was then elaborated in Sammis et al. (1996), Saleur et al. (1996a, b), Johansen et al. (1996), and Huang et al. (1998). The idea was tested experimentally in Bowman et al. (1998).

"On Monday, October 27, 1997 . . .": These numbers are from the U.S. Securities and Exchange Commission (1998).

171 *"Sornette and Ledoit, however, made a 400% profit"*: See Sornette (2003, p. 250).

"Historians now explain the worldwide crash . . .": See, for instance, Radelet and Sachs (2000).

172 *"Critical phenomena often have . . ."*: See, for instance, Batterman (2002) and

references therein for a detailed account of the role of universality in the study of critical phenomena.

173 *"Sornette's first foray into economics . . .":* The first paper with Bouchaud was Bouchaud and Sornette (1994).

". . . realized that Sornette's earlier work . . .": The first paper on this topic was Sornette (1996); it was greatly expanded the following year in Sornette and Johansen (1997).

174 *". . . in 1841, Charles Mackay wrote a book . . .":* This is Mackay (1841).

"Perhaps the most striking example . . .": For more on tulip mania, see Dash (1999) and Goldgar (2007); for another, more skeptical perspective, see Thompson (2007).

175 *"That winter, a single bulb . . .":* These numbers are from Dash (1999).

176 *"Since first predicting the October 1997 crash . . .":* See the description of his predictions in Sornette (2003); my reports on his more recent successes are from private communication.

178 *". . . he calls them 'dragon kings'":* See Sornette (2009).

8. A New Manhattan Project

181 *"Pia Malaney put her arms on the table . . .":* Again, this opening section is a dramatization, but the essential facts are correct. Malaney and Weinstein's story has not been told before. The version presented here is told from their point of view—including their speculations about the motivations of some of the principal characters involved—and is based primarily on numerous interviews with Weinstein, as well as an interview with Malaney and an interview with Lee Smolin. Malaney and Weinstein read an earlier draft of the chapter and offered helpful comments on tone and accuracy.

183 *"How much is money worth?":* For background on index numbers, see Mankiw (2012) and Krugman and Wells (2009). For more detailed discussions, see Turvey (2004), Barnett and Chauvet (2010), Handa (2000), or Allen (1975).

185 *"In June 1995 the U.S. Senate appointed . . .":* See Boskin et al. (1996) for the final report, as well as Boskin et al. (1998). Histories of the Boskin Commission can be found in Sheehan (2010), Baker (1998), Baker and Weisbrot (1999), and Gordon (2006). An account of how the Bureau of Labor Statistics responded to the Boskin report can be found in Greenlees (2006).

". . . soon-to-be-disgraced Senator Bob Packwood . . .": Packwood resigned from the Senate on September 7, 1995, under a cloud of alleged sexual misconduct.

186 *"Hermann Weyl was offered the position of chair . . .":* For more on Weyl's biography and contributions to geometry, see Atiyah (2003) and Scholz (1994).

"One of these was a prominent young physicist . . .": For more on Einstein's biography, see Isaacson (2007), Galison (2003), and Pais (1982).

187 *"The basic idea underlying general relativity . . .":* For more on general relativity, see Misner et al. (1973) and Wald (1984). The best nontechnical introduction to the subject is Geroch (1981).

191 *"Weyl called his new theory a gauge theory"*: For more on the history of gauge theory, including Weyl's early contributions, see O'Raifeartaigh (1997).

192 *". . . accelerated when Yang, in collaboration with Jim Simons . . ."*: Simons and Yang describe their collaboration in Zimmerman (2009); see also the famous Wu-Yang dictionary (Wu and Yang 1975).

"This framework was called the Standard Model . . .": For more on the Standard Model, see Hoddeson et al. (1997) for the history and Cottingham and Greenwood (2007) for the physics.

193 *"Afterward, she and Maskin met . . ."*: The version of events described here is due to Weinstein and Malaney. I contacted Maskin to confirm the story, but he indicated that he did not remember the sequence of events clearly enough to comment.

194 *"Jorgenson replied by throwing her out of his office"*: Again, this is Weinstein and Malaney's account. They recall that Jorgenson's criticism of Malaney's thesis project was that (he claimed) she had merely found a new way to derive something called the Divisia index. Divisia indices (or Divisia monetary aggregates) provide an alternative method of measuring economic variables like inflation. Though these methods were already well known (but not widely used) when Malaney presented her ideas to Jorgenson, Malaney and Weinstein's novel approach to deriving the Divisia index had already yielded significant new results. For more on the Divisia index, see Divisia (1925), Barnett and Chauvet (2010), and Handa (2000). In particular, see Barnett (2012), which argues that U.S. monetary policy is plagued by significant problems stemming from the use of inappropriate statistical measures, and in particular from not using Divisia indices more widely. In other words, there is a political question, closely related to the question that the Boskin Commission was set to work on, concerning the Divisia index.

". . . a classic result in economics known as Coase's theorem": Coase's theorem is originally described in Coase (1960). See also Krugman and Wells (2009).

197 *"To solve the index number problem . . ."*: This proposal is described in detail in Malaney's thesis (Malaney 1996) and (in somewhat modified form) in Smolin (2009). See also Illinski (2001), which takes a rather different approach to applying gauge theory in economics, as well as Didier Sornette's criticism of Illinski's book (Sornette 1998).

"One day, late in 2005, Lee Smolin . . .": The story told in this section is based in part on an interview with Smolin.

198 *". . . Smolin had published an article in the magazine* Physics Today *. . ."*: The article was Smolin (2005).

". . . a book Smolin was just finishing . . .": This book was Smolin (2006).

199 *"He gave a talk on the way gauge-theoretic ideas . . ."*: This talk, and Weinstein's talks at the conferences described in the text, are available through Perimeter's online archive. See Weinstein (2006, 2008, 2009).

200 *"Social Security . . . was first signed into law . . ."*: For a history of the politics surrounding Social Security, see Beland (2005), Altman (2005), or Baker and Weisbrot (1999).

201 *". . . until Daniel Patrick Moynihan and Bob Packwood . . . shared a moment of*

inspiration . . .": See Sheehan (2010), Moynihan (1996), and Katzmann (2008) for various perspectives on the origins of the Boskin Commission.

"According to notes written by Robert Gordon . . .": The notes I am alluding to are Gordon (2002). Gordon also tells the story of the Boskin Commission and its critics in Gordon (2006), though this narrative version does not include Moynihan's role.

202 *"The Boskin Commission's findings were criticized from all corners":* See especially Sheehan (2010), as well as Triplett (2006) and Bosworth (1997).

"Ultimately, the Boskin Commission's recommendations were squashed . . .": For an account of how the Bureau of Labor Statistics incorporated some of the Boskin Commission's recommendations, see Greenlees (2006). The NAS report was published as Schultze and Mackie (2002).

203 *"The plan was to use this conference . . .":* For a discussion of Weinstein's idea, with comments by Weinstein, see Brown et al. (2008). See also Weinstein (2009).

1. Epilogue: Send Physics, Math, and Money!

206 *". . . to pen the 'Financial Modelers' Manifesto' ":* This is Derman and Wilmott (2009).

210 *"In the words of sociologist Donald MacKenzie . . .":* I am alluding to the title of MacKenzie (2006), *An Engine, Not a Camera.* MacKenzie's central point here is that financial markets are shaped by the models that we use to understand them. This strikes me as exactly right, and it presents a special difficulty for scientists and mathematicians trying to study markets.

212 *"In a 1965 Supreme Court decision . . .":* This decision was *Lamont v. Postmaster General;* it can be found in Sepinuck and Treuthart (1999, Ch. 2). For more on Brennan's ideas about expression, see the other opinions in that collection or Hopkins (1991).

213 *". . . an argument from psychology and human behavior":* For a sample of this view, see Brooks (2010). It is closely related to arguments concerning behavioral economics, as seen (for instance) in Ariely (2008), Akerlof (2009), or Shiller (2005). For more scholarly work on behavioral finance, one might start with Thaler (1993, 2005). However, as should be clear in the text, I want to distinguish between behavioral economics as a discipline—which has clearly made enormous progress in understanding the psychology and sociology of economic decision making—and a specific argument based on some results of behavioral economics to the effect that mathematical modeling in finance is impossible, or that, as Brooks puts it, economics should be "an art, not a science." It is only the latter argument that I object to here; behavioral economics more generally, I think, plays an essential role in identifying how certain assumptions regarding rational behavior are unrealistic and guides the way in the construction of future models that (one hopes) are able to account for the "predictably irrational" behaviors of real investors.

214 *". . . has found its biggest champion in Nassim Taleb":* See in particular Taleb (2004, 2007a). But see also Taleb (2007b), where (it seems to me) Taleb is more moderate.

216 "*. . . Warren Buffett, who has famously warned . . .*": Buffett appears to express these opinions regularly, but the specific quote about geeks bearing formulas is from Buffett (2008, p. 14).

217 "*. . . it's not as though market crashes . . . are a new phenomenon . . .*": Indeed, financial crisis has plagued us for as long as we've had economies. For excellent histories of financial calamity, see Reinhart and Rogoff (2009) and Kindleberger and Aliber (2005).

218 "*. . . the SEC has charged Goldman Sachs . . .*": For more on the SEC accusations and the settlement reached with Goldman Sachs, see U.S. Securities and Exchange Commission (2010a, b).

219 "*. . . in Buffett's words, 'financial weapons of mass destruction' . . .*": This quote is from Buffett (2002, p. 15).

220 "*. . . because in 1934 the U.S. government instituted the Federal Deposit Insurance Corporation . . .*": For more on the history of the FDIC, and other changes in the financial regulation in the United States, see Markham (2002).

"*But what caused the quant crisis?*": There remain many views on what caused the financial crisis, including the quant crisis. For instance, see Shiller (2008), Krugman (2008), Zandi (2008), McLean and Nocera (2010), or the Financial Crisis Inquiry Commission (2011). The analysis presented here is heavily indebted to Gorton (2010).

222 "*First was Renaissance . . .*": This part of the story is from Patterson (2010).

"*The securitization procedure by which subprime mortgages . . .*": The model is presented in Li (2000). See also Salmon (2009).

224 "*. . . it seems poised for another collapse*": For a picture of the deep financial and economic problems lurking under the surface of the world's economies, see Rajan (2010).

References

Aelian, Claudius. 1959 (200 A.D.). *On the Characteristics of Animals [De animalium natura]*, ed. A. F. Scholfield. London: Heinemann.

Ahrens, Frank. 2007. "For Wall Street's Math Brains, Miscalculations." *The Washington Post*, August 21.

Akerlof, George A. 2009. *Animal Spirits: How Human Psychology Drives the Economy, and Why It Matters for Global Capitalism.* Princeton, NJ: Princeton University Press.

Allegre, C. J., J. L. Le Moule, and A. Provost. 1982. "Scaling Rules in Rock Fracture and Possible Implications for Earthquake Predictions." *Nature* 297: 47–49.

Allen, R.G.D. 1975. *Index Numbers in Economic Theory and Practice.* Piscataway, NJ: Transaction Publishers.

Allen, Robert Loring. 1993. *Irving Fisher: A Biography.* Cambridge, MA: Wiley-Blackwell.

Allison, David K. 1985. "U.S. Navy Research and Development Since World War II." In *Military Enterprise and Technological Change: Perspectives on the American Experience*, ed. Merritt Roe Smith. Cambridge, MA: MIT Press.

Altman, Nancy. 2005. *The Battle for Social Security: From FDR's Vision to Bush's Gamble.* Hoboken, NJ: John Wiley and Sons.

Anderson, P. W., K. Arrow, and D. Pines. 1988. *The Economy as an Evolving Complex System.* Reading, MA: Addison-Wesley.

Anifrani, J. C. 1995. "Universal Log-Periodic Correction to Renormalization Group Scaling for Rupture Stress Prediction From Acoustic Emissions." *Journal de Physique I* 5 (6): 631.

Ansbacher, Max. 2000. *The New Options Market.* Hoboken, NJ: John Wiley and Sons.

Ariely, Dan. 2008. *Predictably Irrational.* New York: HarperCollins.

Arthur, W. B., S. N. Durlauf, and D. A. Lane, eds. 1997. *The Economy as a Complex Evolving System II.* Reading, MA: Addison-Wesley.

Atiyah, M. 2003. "Hermann Weyl: November 9, 1885–December 9, 1955." *Biographical Memoirs National Academy of Sciences* 82: 320–35.

Bachelier, Louis. 1900. "Théorie de la spéculation." *Annales Scientifiques de l'École Normale Supérieure* 19: 21–86.

———. 1914. *Le jeu, la chance et le hasard.* Paris: Ernest Flammarion.

———. 1937. *Les lois des grands nombres du calcul des probabilités.* Paris: Gauthier-Villars.

———. 1941. "Probabilités des oscillations maxima." *Comptes-rendus hebdomadaires des séances de l'Académie des Sciences,* May, 836–38.

Baggott, Jim. 2009. *Atomic: The First War of Physics and the Secret History of the Atom Bomb: 1939–49.* London: Icon Books.

Bak, P. 1996. *How Nature Works: The Science of Self-Organized Criticality.* New York: Springer-Verlag.

Bak, Per, Chao Tang, and Kurt Wiesenfeld. 1987. "Self-Organized Criticality: An Explanation of the 1/*f* Noise." *Physical Review Letters* 59 (4, July): 381–84.

Baker, Dean, ed. 1998. *Getting Prices Right: The Debate Over the Consumer Price Index.* New York: M. E. Sharpe, Inc.

Baker, Dean, and Mark Weisbrot. 1999. *Social Security: The Phony Crisis.* Chicago: University of Chicago Press.

Baldwin, Roger R., Wilbert E. Cantey, Herbert Maisel, and James P. McDermott. 1956. "The Optimum Strategy in Blackjack." *Journal of the American Statistical Association* 51 (275): 429–39.

Barcellos, Anthony. 1985. "Benoît Mandelbrot." In *Mathematical People,* ed. Donald J. Albers and G. L. Alexanderson. Boston: Birkhäuser.

Barnett, W. A. 2012. *Getting It Wrong: How Faulty Monetary Statistics Undermine the Fed, the Financial System, and the Economy.* Cambridge, MA: MIT Press.

Barnett, William A., and Marcelle Chauvet. 2010. *Financial Aggregation and Index Number Theory.* Singapore: World Scientific Publishing.

Bass, Thomas A. 1985. *The Eudaemonic Pie.* Boston: Houghton Mifflin.

———. 1999. *The Predictors: How a Band of Maverick Physicists Used Chaos Theory to Trade Their Way to a Fortune on Wall Street.* New York: Henry Holt.

Batterman, Robert. 2002. *The Devil in the Details: Asymptotic Reasoning in Explanation, Reduction, and Emergence.* Oxford: Oxford University Press.

Beland, Daniel. 2005. *Social Security: History and Politics From the New Deal to the Privatization Debate.* Lawrence: University Press of Kansas.

Bernstein, Jeremy. 2010. *Physicists on Wall Street and Other Essays on Science and Society.* New York: Springer Business + Media.

Bernstein, Peter. 1993. *Capital Ideas: The Improbable Origins of Modern Wall Street.* New York: Free Press.

———. 1998. *Against the Gods: The Remarkable Story of Risk.* Hoboken, NJ: John Wiley and Sons.

Bhat, M. R. 1981. *Varahamihira's Brhat Samhita.* Delhi: Motilal Banarsidass.

Billingsley, P. 1995. *Probability and Measure.* New York: John Wiley and Sons.

Bird, Kai, and Martin J. Sherwin. 2005. *American Prometheus: The Triumph and Tragedy of Robert Oppenheimer.* New York: Random House.

Black, Fischer. 1987. *Business Cycles and Equilibrium.* Hoboken, NJ: John Wiley and Sons.

———. 1989. "How We Came Up with the Option Formula." *Journal of Portfolio Management* 15 (2): 4–8.

———. 1992. "The Holes in Black-Scholes." In *From Black-Scholes to Black Holes: New Frontiers in Options*, 51–56. London: Risk Magazine.

———. 2010. *Exploring General Equilibrium.* Cambridge, MA: MIT Press.

Black, Fischer, and Myron Scholes. 1972. "The Valuation of Option Contracts and a Test of Market Efficiency." *Journal of Finance* 27 (2): 399–418.

———. 1973. "The Pricing of Options and Corporate Liabilities." *Journal of Political Economy* 81 (3): 637–54.

———. 1974. "From Theory to a New Financial Product." *Journal of Finance* 19 (2): 399–412.

Blume, L. E., and Steven N. Durlauf, eds. 2006. *The Economy as an Evolving Complex System III: Current Perspectives and Future Directions* (Santa Fe Institute Studies in the Science of Complexity). New York: Oxford University Press.

Bookstaber, Richard. 2007. *A Demon of Our Own Design: Markets, Hedge Funds and the Perils of Financial Innovation.* Hoboken, NJ: John Wiley and Sons.

Boskin, Michael J., E. Dullberger, R. Gordon, Z. Griliches, and D. Jorgenson. 1996. "Towards a More Accurate Measure of the Cost of Living." Final Report to the Senate Finance Committee, December 4.

———. 1998. "Consumer Prices, the Consumer Price Index, and the Cost of Living." *Journal of Economic Perspectives* 12 (1, Winter): 3–26.

Bosworth, Barry P. 1997. "The Politics of Immaculate Conception." *The Brookings Review,* June, 43–44.

Bouchaud, Jean-Philippe, and Didier Sornette. 1994. "The Black-Scholes Option Pricing Problem in Mathematical Finance: Generalization and Extensions for a Large Class of Stochastic Processes." *Journal de Physique* 4 (6): 863–81.

Bower, Tom. 1984. *Klaus Barbie, Butcher of Lyons.* London: M. Joseph.

Bowman, D. D., G. Ouillion, C. G. Sammis, A. Sornette, and D. Sornette. 1998. "An Observational Test of the Critical Earthquake Concept." *Journal of Geophysical Research* 103: 24359–72.

Broad, William J. 1992. "Defining the New Plowshares Those Old Swords Will Make." *The New York Times,* February 5.

Brooks, David. 2010. "The Return of History." *The New York Times,* March 26, A27.

Brown, Mike, Stuart Kauffman, Zoe-Vonna Palmrose, and Lee Smolin. 2008. "Can Science Help Solve the Economic Crisis?" Available, with a response from Weinstein, at http://www.edge.org/conversation/can-science-help-solve-the-economic-crisis.

Brown, Robert. 1828. "A Brief Account of Microscopical Observations Made on the Particles Contained in the Pollen of Plants." *Philosophical Magazine* 4: 161–73.

Bruck, Connie. 1994. *Master of the Game: Steve Ross and the Creation of Time Warner.* New York: Simon & Schuster.

Bufe, Charles G., and David J. Varnes. 1993. "Predictive Modeling of the Seismic Cycle of the Greater San Francisco Bay Region." *Journal of Geophysical Research* 98 (B6): 9871–83.

Buffett, Warren. 2002. "Annual Shareholder Letter." Available at http://www.berkshirehathaway.com/letters/2002pdf.pdf.

———. 2008. "Annual Shareholder Letter." Available at http://www.berkshirehathaway.com/letters/2008ltr.pdf.

———. 2010. "Annual Shareholder Letter." Available at http://www.berkshirehathaway.com/letters/2010ltr.pdf.

Cadbury, Deborah. 2006. *Space Race: The Epic Battle Between America and the Soviet Union for Dominion of Space.* New York: HarperCollins.

Cardano, Girolamo. 1929 (1576). *The Book of My Life [De vita propria liber],* trans. Jean Stoner. New York: E. P. Dutton.

———. 1961 (1565). *The Book on Games of Chance [Liber de ludo aleae],* trans. Sydney Henry Gould. New York: Holt, Rinehart and Winston.

Casella, George, and Roger L. Berger. 2002. *Statistical Inference.* 2nd ed. Pacific Grove, CA: Duxbury.

Cassidy, John. 2007. "The Blow-Up Artist." *The New Yorker,* October 15, 56–69.

———. 2010a. "After the Blowup." *The New Yorker,* January 11, 28–33.

———. 2010b. *How Markets Fail.* New York: Farrar, Straus and Giroux.

Cervantes, Miguel de. 1881. *The Exemplary Novels of Cervantes,* ed. Walter K. Kelly. London: George Bell and Sons.

Chalmers, Alan. 2009. *The Scientist's Atom and the Philosopher's Stone: How Science Succeeded and Philosophy Failed to Gain Knowledge of Atoms.* New York: Springer-Verlag.

———. 2011. "Drawing Philosophical Lessons From Perrin's Experiments on Brownian Motion: A Response to van Fraassen." *British Journal of the Philosophy of Science* 62 (4): 711–32.

Chapman, Toby. 1998. "Speculative Trading: Physicists' Forays Into Finance." *Europhysics Notes,* January/February, 4.

Cirillo, Renato. 1979. *The Economics of Vilfredo Pareto.* New York: Frank Cass and Company.

Coase, Ronald H. 1960. "The Problem of Social Cost." *Journal of Law and Economics III,* October, 1–44.

Cole, K. C. 2009. *Something Incredibly Wonderful Happens: Frank Oppenheimer and the World He Made Up.* Boston: Houghton Mifflin Harcourt.

Collins, Martin. 1999. *Space Race: The U.S.-U.S.S.R. Competition to Reach the Moon.* Rohnert Park, CA: Pomegranate Communications.

Compton, Arthur Holly. 1956. *Atomic Quest.* New York: Oxford University Press.

Conant, Jennet. 2005. *109 East Palace: Robert Oppenheimer and the Secret City of Los Alamos.* New York: Simon & Schuster.

Cone, Edward. 1999. "Got Risk?" *Wired* 7 (12).

Cont, R. 2001. "Empirical Properties of Asset Returns: Stylized Facts and Statistical Issues." *Quantitative Finance* 1: 223–36.

Cootner, Paul, ed. 1964. *The Random Character of Stock Prices.* Cambridge, MA: MIT Press.

Cottingham, W. N., and D. A. Greenwood. 2007. *An Introduction to the Standard Model of Particle Physics.* Cambridge: Cambridge University Press.

Courtault, Jean-Michel, and Youri Kabanov. 2002. *Louis Bachelier: Aux origines de la finance mathématique.* Paris: Presses Universitaires Franc-Comtoises.

Cox, John C., and Mark Rubinstein. 1985. *Options Markets.* Englewood Cliffs, NJ: Prentice Hall.

Dash, Mike. 1999. *Tulipomania: The Story of the World's Most Coveted Flower and the Extraordinary Passions It Aroused.* New York: Three Rivers Press.

David, F. N. 1962. *Games, Gods & Gambling: A History of Probability and Statistical Ideas.* New York: Simon & Schuster.

Davis, Mark, and Alison Etheridge. 2006. *Louis Bachelier's Theory of Speculation: The Origins of Modern Finance.* Princeton, NJ: Princeton University Press.

Davis, Monte. 1984. "Benoît Mandelbrot." *Omni Magazine* 6 (5): 64.

Davy, P. H., A. Sornette, and D. Sornette. 1990. "Some Consequences of a Proposed Fractal Nature of Continental Faulting." *Nature* 348 (November): 56–58.

Derman, Emanuel. 2004. *My Life as a Quant.* Hoboken, NJ: John Wiley and Sons.

———. 2011a. "Emanuel Derman on Fischer Black." Available at https://www.quantnet.com/emanuel-derman-fischer-black/.

———. 2011b. *Models Behaving Badly.* New York: Free Press.

Derman, Emanuel, and Iraj Kani. 1994. "The Volatility Smile and Its Implied Tree." Goldman Sachs Quantitative Strategies Research Note.

Derman, Emanuel, and Nassim Nicholas Taleb. 2005. "The Illusions of Dynamic Replication." *Quantitative Finance* (4): 323–26.

Derman, Emanuel, and Paul Wilmott. 2009. "The Financial Modelers' Manifesto." Available at Social Science Research Network (SSRN), http://ssrn.com/abstract=1324878 or http://dx.doi.org/10.2139/ssrn.1324878.

Devlin, Keith. 2008. *The Unfinished Game: Pascal, Fermat, and the Seventeenth-Century Letter That Made the World Modern.* New York: Basic Books.

Dimand, Robert W., and Hichem Ben-El-Mechaiekh. 2006. "Louis Bachelier." In *Pioneers of Financial Economics*, vol. 1, ed. Geoffrey Poitras. Northampton, MA: Edward Elgar Publishing.

Divisia, François. 1925. "L'Indice monétaire et la théorie de la monnaie." *Revue d'Économie Politique* 3: 842–64.

Duffus, R. L. 1972. *The Santa Fe Trail.* Albuquerque: University of New Mexico Press.

Dwork, Deborah, and Robert Jan van Pelt. 2002. *Holocaust: A History.* New York: W. W. Norton.

Eichengreen, Barry. 2008. *Globalizing Capital: A History of the International Monetary System.* 2nd ed. Princeton, NJ: Princeton University Press.

Eichenwald, Kurt. 1989a. "Jury Selection Begins Today in Princeton/Newport Case." *The New York Times,* June 19.

———. 1989b. "Six Guilty of Stock Conspiracy." *The New York Times,* August 1.

Einstein, Albert. 1905a. "Über einen die Erzeugung und Verwandlung des Lichtes betreffenden heuristischen Gesichtspunkt." *Annalen der Physik* 17: 132–48.

———. 1905b. "Über die von der molekularkinetischen Theorie der Wärme geforderte Bewegung von in ruhenden Flüssigkeiten suspendierten Teilchen." *Annalen der Physik* 17: 549–60.

———. 1905c. "Zur Elektrodynamik bewegter Körper." *Annalen der Physik* 17: 891–921.

———. 1905d. "Ist die Trägheit eines Körpers von seinem Energiegehalt abhängig?" *Annalen der Physik* 18, 639–41.

———. 1946. *The Meaning of Relativity.* 2nd ed. Princeton, NJ: Princeton University Press.

Falconer, Kenneth. 2003. *Fractal Geometry: Mathematical Foundations and Applications.* 2nd ed. Hoboken, NJ: John Wiley and Sons.

Fama, Eugene. 1964. "Mandelbrot and the Stable Paretian Hypothesis." In *The Random Character of Stock Prices,* ed. Paul Cootner, 297–306. Cambridge, MA: MIT Press.

———. 1965. "The Behavior of Stock Market Prices." *Journal of Business* 38 (1).

Farmer, J. Doyne, and John J. Sidorowich. 1987. "Predicting Chaotic Time Series." *Physical Review Letters* 59 (8): 845–48.

Figlewski, Stephen. 1995. "Remembering Fischer Black." *The Journal of Derivatives* 3 (2): 94–98.

Financial Crisis Inquiry Commission. 2011. *The Financial Crisis Inquiry Report, Authorized Edition: Final Report of the National Commission on the Causes of the Financial and Economic Crisis in the United States.* New York: Public Affairs.

Fischel, Jack R. 1998. *The Holocaust.* Westport, CT: Greenwood Press.

Forbes, Catherine, Merran Evans, Nicholas Hastings, and Brian Peacock. 2011. *Statistical Distributions.* 4th ed. Hoboken, NJ: John Wiley and Sons.

Forbes magazine. 2011. "The World's Billionaires 2011." Available at http://www.forbes.com/lists/2011/10/billionaires_2011.html.

Forfar, David O. 2007. "Fischer Black." Available at http://www.history.mcs.standrews.ac.uk/Biographies/Black_Fischer.html.

Fox, Justin. 2009. *The Myth of the Rational Market.* New York: Harper Business.

French, Craig W. 2003. "The Treynor Capital Asset Pricing Model." *Journal of Investment Management* 1 (2): 60–72.

Galison, Peter. 1997. *Image and Logic: A Material Culture of Microphysics.* Chicago: University of Chicago Press.

———. 2003. *Einstein's Clocks, Poincaré's Maps: Empires of Time.* New York: W. W. Norton.

Galison, Peter, and Bruce Hevly, eds. 1992. *Big Science.* Stanford, CA: Stanford University Press.

Gebhard, Louis A. 1979. *Evolution of Naval Radio-Electronics and Contributions of the Naval Research Laboratory.* Washington, DC: Naval Research Laboratory. NRL Report 8300.

Geroch, Robert. 1981. *General Relativity From A to B.* Chicago: University of Chicago Press.

Girlich, Hans-Joachim. 2002. "Bachelier's Predecessors." Available at http://www.mathematik.uni-leipzig.de/preprint/2002/p5-2002.pdf.

Glansdorff, Paul, and Ilya Prigogine. 1971. *Thermodynamic Theory of Structure, Stability and Fluctuations.* London: Wiley Interscience.

Gleick, J. 1987. *Chaos: Making a New Science.* New York: Viking.

———. 2011. *The Information: A History, a Theory, a Flood.* Toronto: Pantheon Books.

Goldgar, Anne. 2007. *Tulipmania: Money, Honor, and Knowledge in the Dutch Golden Age.* Chicago: University of Chicago Press.

Gordon, Robert J. 2002. "The Boskin Report vs. NAS *At What Price*: 'The Wild vs. the Mild.'" Slides presented at the 2002 Conference on Research in Income and Wealth. Available at http://faculty-web.at.northwestern.edu/economics/gordon/BoskinvsNAS.ppt.

———. 2006. "The Boskin Commission Report: A Retrospective One Decade Later." *International Productivity Monitor* 12 (June): 7–22.

Gorton, Gary. 2010. *Slapped by the Invisible Hand: The Panic of 2007.* New York: Oxford University Press.

Gray, Robert M. 2011. *Entropy and Information Theory.* New York: Springer-Verlag.

Greenlees, John S. 2006. "The BLS Response to the Boskin Commission Report." *International Productivity Monitor* 12 (June): 23–41.

Greer, John F., Jr. 1996. "Simons Doesn't Say." *Financial World*, October 21.

Groves, Leslie R. 1962. *Now It Can Be Told.* New York: Harper & Row.

Guggenheimer, J., and P. Holmes. 1983. *Nonlinear Oscillations, Dynamical Systems, and Bifurcation of Vector Fields.* Berlin: Springer-Verlag.

Hacking, Ian. 1975. *The Emergence of Probability.* New York: Cambridge University Press.

———. 1990. *The Taming of Chance.* New York: Cambridge University Press.

Hájek, Alan. 2012. "Interpretations of Probability." *The Stanford Encyclopedia of Philosophy,* Spring 2012 edition, ed. Edward N. Zalta. Palo Alto, CA: Center for the Study of Language and Information. Available at http://plato.stanford.edu/archives/spr2012/entries/probability-interpret/.

Hald, Anders. 2003. *A History of Probability and Statistics and Their Applications Before 1750.* Hoboken, NJ: John Wiley and Sons.

Handa, Jagdish. 2000. *Monetary Economics.* New York: Taylor and Francis.

Handley, Susannah. 2000. *Nylon: The Story of a Fashion Revolution.* Baltimore, MD: Johns Hopkins University Press.

Hastings, Charles Sheldon. 1909. *Biographical Memoir of Josiah Willard Gibbs, 1879–1903.* Washington, DC: National Academy of Sciences.

Haug, Espen Gaarder, and Nassim Nicholas Taleb. 2011. "Option Traders Use (Very) Sophisticated Heuristics, Never the Black-Scholes-Merton Formula." *Journal of Economic Behavior and Organization* 77 (2): 97–106.

Hendry, David F., and Mary S. Morgan. 1996. "Obituary: Jan Tinbergen, 1903–1994." *Journal of the Royal Statistics Society: Series A* 159 (3): 614–18.

Hoddeson, Lillian, Laurie Brown, Michael Riordan, and Max Dresden. 1997. *The Rise of the Standard Model: Particle Physics in the 1960s and 1970s.* Cambridge: Cambridge University Press.

Hopkins, W. Wat. 1991. *Mr. Justice Brennan and Freedom of Expression.* New York: Praeger Publishers.

Hounshell, David A. 1992. "Du Pont and the Management of Large-Scale Research and Development." In *Big Science,* ed. Peter Galison and Bruce Hevly. Stanford, CA: Stanford University Press.

Hounshell, David A., and John Kenly Smith Jr. 1988. *Science and Corporate Strategy: Du Pont R&D, 1902–1980.* New York: Cambridge University Press.

Huang, Y., H. Saleur, C. Sammis, and D. Sornette. 1998. "Precursors, Aftershocks, Criticality and Self-Organized Criticality." *Europhysics Letters* 41: 44–48.

Hull, John C. 2011. *Options, Futures, and Other Derivatives.* 8th ed. Upper Saddle River, NJ: Prentice Hall.

Hunsaker, Jerome, and Saunders MacLane. 1973. *Edwin Bidwell Wilson: 1879–1964.* Washington, DC: National Academy of Sciences.

Illinski, K. 2001. *The Physics of Finance: Gauge Modelling in Non-equilibrium Pricing.* New York: John Wiley and Sons.

Isaacson, Walter. 2007. *Einstein: His Life and Universe.* New York: Simon & Schuster.

Johansen, A., and D. Sornette. 2000. "Critical Ruptures." *The European Physical Journal B—Condensed Matter and Complex Systems* 18 (1): 163–81.

Johansen, Anders, Didier Sornette, Hiroshi Wakita, Urumu Tsunogai, William I. Newman, and Hubert Saleur. 1996. "Discrete Scaling in Earthquake Precursory Phenomena: Evidence in the Kobe Earthquake, Japan." *Journal de Physique I* 6 (10): 1391–1402.

Jones, Vincent C. 1985. *Manhattan, the Army and the Atomic Bomb.* Washington, DC: Government Printing Office.

Jovanovic, Frank. 2000. "L'origine de la théorie financière: Une réévaluation de l'apport de Louis Bachelier." *Revue d'Économie Politique* 110 (3): 395–418.

———. 2006. "A Nineteenth-Century Random Walk: Jules Regnault and the Origins of Scientific Financial Economics." In *Pioneers of Financial Economics,* vol. 1, ed. Geoffrey Poitras. Northampton, MA: Edward Elgar Publishing.

Jung, Jayne. 2007. "The Right Time." *Risk Magazine,* September 1.

Kahn, David. 1967. *The Code-Breakers: The Comprehensive History of Secret Communication From Ancient Times to the Internet.* New York: Scribner.

Kaplan, Ian. 2002. "*The Predictors* by Thomas A. Bass: A Retrospective." This is a comment on *The Predictors* by a former employee of the Prediction Company. Available at http://www.bearcave.com/bookrev/predictors2.html.

Karlin, Samuel, and Howard M. Taylor. 1975. *A First Course in Stochastic Processes.* 2nd ed. San Diego, CA: Academic Press.

———. 1981. *A Second Course in Stochastic Processes.* San Diego, CA: Academic Press.

Katzmann, Robert A. 2008. *Daniel Patrick Moynihan: The Intellectual in Public Life.* Washington, DC: Woodrow Wilson Center Press.

Kelly, J., Jr. 1956. "A New Interpretation of Information Rate." *IRE Transactions on Information Theory* 2 (3, September): 185–89.

Kelly, Kevin. 1994a. "Cracking Wall Street." *Wired* 2 (7).

———. 1994b. *Out of Control: The Rise of Neobiological Civilization.* Reading, MA: Addison-Wesley.

Kendall, M. G. 1953. "The Analysis of Economic Time-Series, Part 1: Prices." *Journal of the Royal Statistical Society* 116 (1): 11–34.

Khandani, Amir E., and Andrew W. Lo. 2011. "What Happened to the Quants in August 2007? Evidence From Factors and Transactions Data." *Journal of Financial Markets* 14 (1): 1–46.

Kindleberger, Charles P., and Robert Aliber. 2005. *Manias, Panics, and Crashes.* Hoboken, NJ: John Wiley and Sons.

Kolmogorov, Andrei. 1931. "Über die analytischen Methoden in der Wahrscheinlichkeitsrechnung." *Mathematische Annalen* 104: 415–58.

Krige, J., A. Russo, and L. Sebesta. 2000. *The Story of ESA, 1973–1987,* vol. 2 of *A History of the European Space Agency, 1958–1987.* Noordwijk: ESA Publications Division.

Krugman, Paul. 2008. *The Return of Depression Economics and the Crisis of 2008.* New York: W. W. Norton.

———. 2009. "How Did Economists Get It So Wrong?" *The New York Times Magazine,* September 6.

Krugman, Paul, and Robin Wells. 2009. *Economics.* 2nd ed. New York: Worth Publishers.

Lahart, Justin. 2007. "Behind the Stock Market's Zigzag." *The Wall Street Journal,* August 11, B1.

Laing, Jonathan R. 1974. "Playing the Odds." *The Wall Street Journal,* September 23, 1.

Lamaignère, Laurent, François Carmona, and Didier Sornette. 1996. "Experimental Realization of Critical Thermal Fuse Rupture." *Physical Review Letters* 77 (13, September): 2738–41.

———. 1997. "Static and Dynamic Electrical Breakdown in Conducting Filled-Polymers." *Physica A: Statistical Mechanics and Its Applications* 241 (1–2): 328–33.

Lehmann, Bruce N., ed. 2005. *The Legacy of Fischer Black.* New York: Oxford University Press.

Lehmann, P. J. 1991. *La Bourse de Paris.* Paris: Dunod.

———. 1997. *Histoire de la Bourse de Paris.* Paris: Presses Universitaires France.

Li, David X. 2000. "On Default Correlation: A Copula Function Approach." *Journal of Fixed Income* 9 (4): 43–54.

Li, Tien-Yien, and James A. Yorke. 1975. "Period Three Implies Chaos." *The American Mathematical Monthly* 82 (10): 985–92.

Lim, Kian-Guan. 2006. "The Efficient Market Hypothesis: A Developmental Perspective." In *Pioneers of Financial Economics,* vol. 2., ed. Geoffrey Poitras. Northampton, MA: Edward Elgar Publishing.

Lintner, John. 1965. "The Valuation of Risk Assets and the Selection of Risky Investments in Stock Portfolios and Capital Budgets." *Review of Economics and Statistics* 47: 13–37.

Lorenz, Edward. 1993. *The Essence of Chaos.* Seattle: University of Washington Press.

———. 2000. "Predictability: Does the Flap of a Butterfly's Wings in Brazil Set Off a Tornado in Texas?" In *The Chaos Avant-Garde: Memories of the Early Days of Chaos Theory,* ed. Ralph Abraham and Yoshisuke Ueda. Singapore: World Scientific Publishing.

Lowenstein, Roger. 2000. *When Genius Failed: The Rise and Fall of Long-Term Capital Management.* New York: Random House.

Lucretius. 2008 (60 B.C.). *Nature of Things [De rerum natura],* trans. David R. Slavitt. Berkeley, CA: University of California Press.

Lux, Hal. 2000. "How Does This Prize-Winning Mathematician and Former Code Breaker Rack Up His Astonishing Returns? Try a Little Luck and a Firm Full of Ph.D.s." *Institutional Investor,* November 1.

Mackay, Charles. 1841. *Extraordinary Popular Delusions and the Madness of Crowds.* London: Richard Bentley.

MacKenzie, Donald. 2006. *An Engine, Not a Camera.* Cambridge, MA: MIT Press.

MacLean, Leonard C., Edward O. Thorp, and William T. Ziemba. 2011. *The Kelly Capital Growth Investment Criterion.* Singapore: World Scientific Publishing.

Maddy, Penelope. 1997. *Naturalism in Mathematics.* New York: Oxford University Press.

———. 2001. "Naturalism: Friends and Foes." *Philosophical Perspectives* 15: 37–67.

———. 2007. *Second Philosophy.* New York: Oxford University Press.

Mahwin, Jean. 2005. "Henri Poincaré. A Life in the Service of Science." *Notices of the AMS* 52 (9): 1036–44.

Malaney, Pia. 1996. "The Index Number Problem: A Differential Geometric Approach." Dissertation defended at Harvard University.

Malevergne, Y., and D. Sornette. 2006. *Extreme Financial Risks: From Dependence to Risk Management.* Berlin: Springer-Verlag.

Malkiel, Burton G. 1973. *A Random Walk Down Wall Street: The Best Investment Advice for the New Century.* New York: W. W. Norton.

Mallaby, Sebastian. 2010. *More Money Than God: Hedge Funds and the Making of a New Elite.* New York: Penguin Press.

Mandelbrot, Benoît. 1964. "The Variation of Certain Speculative Prices." *The Random Character of Stock Prices,* ed. Paul Cootner, 307–32. Cambridge, MA: MIT Press.

———. 1967. "How Long Is the Coast of Britain? Statistical Self-Similarity and Fractional Dimension." *Science* 156 (3775) : 636–38.

———. 1975. *Les objets fractals: Forme, hasard et dimension.* Paris: Flammarion.

———. 1977. *Fractals: Form, Chance, and Dimension.* San Francisco: W. H. Freeman.

———. 1982. *Fractal Geometry of Nature.* New York: W. H. Freeman.

———. 1987. "Exiles in Pursuit of Beauty." *The Scientist,* March 23, 19.

———. 1997. *Fractals and Scaling in Finance: Discontinuity, Concentration, Risk.* New York: Springer-Verlag.

———. 1998. "Personal Narrative Recorded by Web of Stories." Video available at http://www.webofstories.com/play/9596.

———. 2004a. "A Maverick's Apprenticeship." In *The Wolf Prize in Physics,* ed. David Thouless. Singapore: World Scientific Publishing.

———. 2004b. *Fractals and Chaos: The Mandelbrot Set and Beyond.* New York: Springer-Verlag.

———. 2010. "Interview with bigthink.com." Video available at http://bigthink.com/ideas/19207.

Mandelbrot, Benoît, and Richard L. Hudson. 2004. *The Misbehavior of Markets.* New York: Basic Books.

Mankiw, Gregory. 2012. *Principles of Economics.* 6th ed. Mason, OH: South-Western, Cengage Learning.

Mantegna, Rosario N., and H. Eugene Stanley. 2000. *An Introduction to Econophysics: Correlations and Complexity in Finance.* New York: Cambridge University Press.

Markham, Jerry W. 2002. *A Financial History of the United States.* Armonk, NY: M. E. Sharpe.

Marrus, Michael R., and Robert O. Paxton. 1995. *Vichy France and the Jews.* Stanford, CA: Stanford University Press.

McKale, Donald M. 2012. *Nazis After Hitler: How Perpetrators of the Holocaust Cheated Justice and Truth.* Plymouth, UK: Rowman & Littlefield.

McLean, Bethany, and Joe Nocera. 2010. *All the Devils Are Here: The Hidden History of the Financial Crisis.* New York: Portfolio/Penguin.

Mehrling, Perry. 2005. *Fischer Black and the Revolutionary Idea of Finance.* Hoboken, NJ: John Wiley and Sons.

Melamed, Leo. 1993. *Leo Melamed on the Markets.* New York: John Wiley and Sons.

Merton, Robert C. 1973. "Theory of Rational Option Pricing." *Bell Journal of Economics and Management Science* 4 (1): 141–83.

Merton, Robert C., and Myron S. Scholes. 1995. "Fischer Black." *Journal of Finance* 50 (5): 1359–70.

Michie, Ranald C. 1999. *The London Stock Exchange: A History.* New York: Oxford University Press.

Mishkin, Frederic S., and Stanley G. Eakins. 2009. *Financial Markets and Institutions.* 6th ed. Boston, MA: Pearson Education.

Misner, Charles W., Kip S. Thorne, and John Archibald Wheeler. 1973. *Gravitation.* New York: W. H. Freeman.

Mitchell, Melanie. 1998. *An Introduction to Genetic Algorithms.* Cambridge, MA: MIT Press.

Morehead, Albert H. 1967. *Complete Guide to Winning Poker.* New York: Simon & Schuster.

Morgan, Mary S. 1990. *The History of Econometric Ideas.* New York: Cambridge University Press.

———. 2003. "Economics." In *The Cambridge History of Science,* 275–305. New York: Cambridge University Press.

Morley, Henry. 1854. *The Life of Girolamo Cardano, of Milan, Physician.* London: Chapman and Hall.

Moynihan, Daniel P. 1996. *Miles to Go: A Personal History of Social Policy.* Cambridge, MA: Harvard University Press.

Nasar, Sylvia. 1998. *A Beautiful Mind: The Life of Mathematical Genius and Nobel Laureate John Nash.* New York: Touchstone.

Ndiaye, Pap A. 2007. *Nylon and Bombs.* Baltimore, MD: Johns Hopkins University Press.

Niederhoffer, Victor. 1998. *The Education of a Speculator.* Hoboken, NJ: John Wiley and Sons.

Niederhoffer, Victor, and M.F.M. Osborne. 1966. "Market Making and Reversals on the Stock Exchange." *Journal of the American Statistical Association* 61 (316): 897–916.

Nocera, Joe. 2007. "Markets Quake, and a Neutral Strategy Slips." *The New York Times,* August 18, C1.

O'Connor, J. J., and E. F. Robertson. 2005. "Szolem Mandelbrojt." Available at http://www-history.mcs.st-andrews.ac.uk/Biographies/Mandelbrojt.html.

O'Raifeartaigh, Lochlann. 1997. *Dawning of Gauge Theory.* Princeton, NJ: Princeton University Press.

Ore, Øystein. 1953. *Cardano, the Gambling Scholar.* Princeton, NJ: Princeton University Press.

Oreskes, N., and H. Le Grand. 2003. *Plate Tectonics: An Insider's History of the Modern Theory of the Earth.* 2nd ed. Boulder, CO: Westview Press.

Osborne, M.F.M. 1951. "Aerodynamics of Flapping Flight, with Applications to Insects." *Journal of Experimental Biology* 28 (2): 221–45.

———. 1959. "Brownian Motion in the Stock Market." *Operations Research* 7: 145–73.

———. 1961. "The Hydrodynamical Performance of Migratory Salmon." *Journal of Experimental Biology* 38: 365–90.

———. 1962. "Periodic Structure in the Brownian Motion of Stock Prices." *Operations Research* 10 (3): 345–79.

———. 1967. "Some Quantitative Tests for Stock Price Generating Mechanisms and Trading Folklore." *Journal of the American Statistical Association* 62 (318): 321–40.

———. 1973. "The Observation and Theory of Fluctuation in Deep Ocean Currents." *Ergänzungsheft zur Deutschen Hydrographischen Zeitschrift* 8 (13): 1–58.

———. 1977. *The Stock Market and Finance From a Physicist's Viewpoint.* Minneapolis, MN: Crossgar Press.

———. 1987a. "Autobiographical Recollections of M. F. Maury Osborne." Courtesy of the Osborne family.

———. 1987b. "Osborne Family History: Recollections of M.F.M. Osborne." Courtesy of the Osborne family.

Osborne, M.F.M., and Albert Einstein. 1946. Unpublished correspondence. Courtesy of the Osborne family.

Packard, N. H. 1988. "Adaptation Toward the Edge of Chaos." *Dynamic Patterns in Complex Systems,* ed. J.A.S. Kelso, A. J. Mandell, and M. F. Shlesinger. Singapore: World Scientific Publishing.

———. 1990. "A Genetic Learning Algorithm for the Analysis of Complex Data." *Complex Systems* 4 (5): 543–72.

Packard, N. H., J. P. Crutchfield, J. D. Farmer, and R. S. Shaw. 1980. "Geometry From a Time Series." *Physical Review Letters* 45 (9): 712–16.

Pais, Abraham. 1982. *Subtle Is the Lord: The Science and Life of Albert Einstein.* Oxford: Oxford University Press.

———. 2006. *J. Robert Oppenheimer: A Life.* New York: Oxford University Press.

Patterson, Scott. 2010. *The Quants.* New York: Crown Business.

Patterson, Scott, and Anita Raghavan. 2007. "How Market Turmoil Waylaid the 'Quants.'" *The Wall Street Journal,* September 7, A1.

Paxton, Robert O. 1972. *Vichy France: Old Guard and New Order, 1940–1944.* New York: Knopf.

Peltz, Michael. 2008. "James Simons." *Absolute Return + Alpha,* June 20.

Poitras, Geoffrey. 2006. *Pioneers of Financial Economics,* vol. 1. Northampton, MA: Edward Elgar Publishing.

———. 2009. "The Early History of Option Contracts." In *Vinzenz Bronzin's Option Pricing Models,* 487–518. Berlin: Springer-Verlag.

Poundstone, William. 2005. *Fortune's Formula: The Untold Story of the Scientific Betting System That Beat the Casinos and Wall Street.* New York: Hill and Wang.

Poznanski, Renée. 2001. *Jews in France During World War II,* trans. Nathan Bracher. Hanover, NH: Brandeis University Press.

Prigogine, I., and G. Nicolis. 1977. *Self-Organization in Nonequilibrium Systems.* New York: John Wiley and Sons.

Pynchon, Thomas. 1973. *Gravity's Rainbow.* New York: Viking Press.

Radelet, Steven, and Jeffrey D. Sachs. 2000. "The Onset of the East Asian Financial Crisis." In *Currency Crises,* ed. Paul Krugman, 105–62. Chicago: University of Chicago Press.

Rajan, Raghuram G. 2010. *Faultlines.* Princeton, NJ: Princeton University Press.

Reinhart, Carmen M., and Kenneth Rogoff. 2009. *This Time Is Different: Eight Centuries of Financial Folly.* Princeton, NJ: Princeton University Press.

Rhodes, Richard. 1995. *The Making of the Atomic Bomb.* New York: Simon & Schuster.

Rogers, Simon. 2010. "NASA Budgets: US Spending on Space Travel Since 1958." *The Guardian,* February 1. Available at http://www.guardian.co.uk/news/datablog/2010/feb/01/nasa-budgets-us-spending-space-travel.

Rossel, Seymour. 1992. *The Holocaust: The World and the Jews, 1933–1945*. Springfield, NJ: Behrman House.

Rukeyser, Muriel. 1988. *Willard Gibbs*. Woodbridge, CT: Ox Bow Press.

Saichev, Alexander, Yannick Malevergne, and Didier Sornette. 2010. *Theory of Zipf's Law and Beyond*. Berlin: Springer-Verlag.

Saleur, H., C. G. Sammis, and D. Sornette. 1996a. "Discrete Scale Invariance, Complex Fractal Dimensions, and Log-Periodic Fluctuations in Seismicity." *Journal of Geophysical Research* 101 (B8): 17661–77.

———. 1996b. "Renormalization Group Theory of Earthquakes." *Nonlinear Processes in Geophysics* 3 (2): 102–9.

Salmon, Felix. 2009. "Recipe for Disaster: The Formula That Killed Wall Street." *Wired*, 17 (3, October).

Sammis, C. G., D. Sornette, and H. Saleur. 1996. "Complexity and Earthquake Forecasting." In *Reduction and Predictability of Natural Disasters*, ed. J. B. Rundle, W. Klein, and D. L. Turcotte, 143–56. Reading, MA: Addison-Wesley.

Samuelson, Paul. 1947. *Foundations of Economic Analysis*. Cambridge, MA: Harvard University Press.

———. 1948. *Economics*. New York: McGraw-Hill.

———. 2000. "Modern Finance Theory Within One Lifetime." In *Mathematical Finance: Bachelier Congress 2000*, ed. Helyette Geman, Dilip Madan, Stanley R. Pliska, and Ton Vorst. Berlin: Springer-Verlag.

Sauron, Anne Sornette. 1990. "Lois d'echelle dans les milieux fissures: Application à la lithosphere." Dissertation defended at University of Paris-11.

Scholz, Erhard. 1994. "Hermann Weyl's Contributions to Geometry in the Years 1918 to 1923." In *The Intersection of History and Mathematics*, ed. J. Dauben, S. Mitsuo, and C. Saski. Basel: Birkhäuser.

Schultze, Charles, and Christopher Mackie, eds. 2002. *At What Price? Conceptualizing and Measuring Cost-of-Living and Price Indexes*. Washington, DC: National Academies Press.

Schwager, Jack D. 2012. *Market Wizards: Interviews with Top Traders*. Hoboken, NJ: John Wiley and Sons.

Seed magazine. 2006. "James Simons: The Billionaire Hedge Fund Manager Discusses the Impact of Mathematics on His Former Life in Academia and His New One in Finance." September 19.

Sepinuck, Stephen L., and Mary Pat Treuthart, eds. 1999. *The Conscience of the Court: Selected Opinions of Justice William J. Brennan Jr. on Freedom and Equality*. Carbondale: Southern Illinois University Press.

Sewell, Martin. 2011. "A History of the Efficient Market Hypothesis." University College London Department of Computer Science Research Note. Available at http://www-typo3.cs.ucl.ac.uk/fileadmin/UCL-CS/images/Research_Student_Information/RN_11_04.pdf.

Shannon, Claude Elwood, and Warren Weaver. 1949. *A Mathematical Theory of Communication*. Champaign: University of Illinois Press.

Sharpe, William. 1964. "Capital Asset Prices: A Theory of Market Equilibrium Under Conditions of Risk." *Journal of Finance* 19 (3): 425–42.

Sheehan, Frederick J. 2010. *Panderer to Power: The Untold Story of How Alan Greenspan Enriched Wall Street and Left a Legacy of Recession.* New York: McGraw-Hill.

Shiller, Robert J. 2005. *Irrational Exuberance.* 2nd ed. Princeton, NJ: Princeton University Press.

———. 2008. *The Subprime Solution: How Today's Global Financial Crisis Happened, and What to Do About It.* Princeton, NJ: Princeton University Press.

Simons, James. 2010. "Mathematics, Common Sense, and Good Luck: My Life and Careers." A talk delivered at MIT on December 9. Video available at http://video.mit.edu/watch/mathematics-common-sense-and-good-luck-my-life-and-careers-9644.

Siraisi, Nancy G. 1997. *The Clock and the Mirror: Girolamo Cardano and Renaissance Medicine.* Princeton, NJ: Princeton University Press.

Skyrms, Brian. 1999. *Choice and Chance.* 4th ed. Belmont, CA: Wadsworth.

Smalley, R. F., Jr., and D. L. Turcotte. 1985. "A Renormalization Group Approach to the Stick-Slip Behavior of Faults." *Journal of Geophysical Research* 90 (B2, February): 1894–1900.

Smolin, Lee. 2005. "Why No 'New Einstein'?" *Physics Today* 6: 56–57.

———. 2006. *The Trouble with Physics: The Rise of String Theory, the Fall of a Science, and What Comes Next.* New York: Houghton Mifflin.

———. 2009. "Time and Symmetry in Models of Economic Markets." Available at http://arxiv.org/abs/0902.4274.

Sornette, A., and D. Sornette. 1990. "Earthquake Rupture as a Critical Point: Consequences for Telluric Precursors." *Tectonophysics* 179 (34): 327–34.

———. 1996. "Self-Organized Criticality and Earthquakes." *Journal de Physique I* 6: 167–75.

Sornette, Didier. 1996. "Stock Market Crashes Precursors and Replicas." *Journal de Physique I* 6: 167–75.

———. 1998. "Gauge Theory of Finance?" *International Journal of Modern Physics* 9 (3): 505–8.

———. 2000. *Critical Phenomena in Natural Sciences: Chaos, Fractals, Self-Organization and Disorder: Concepts and Tools.* Berlin: Springer-Verlag.

———. 2003. *Why Stock Markets Crash: Critical Events in Complex Financial Systems.* Princeton, NJ: Princeton University Press.

———. 2009. "Dragon Kings, Black Swans and the Prediction of Crises." *International Journal of Terraspace Science Engineering* 2 (1): 1–18.

Sornette, A., P. Davy, and D. Sornette. 1990a. "Growth of Fractal Fault Patterns." *Physical Review Letters* 65 (18, October): 2266–69.

———. 1990b. "Structuration of the Lithosphere in Plate Tectonics as a Self-Organized Critical Phenomenon." *Journal of Geophysical Research* 95 (B11): 17353–61.

Sornette, Didier, and Anders Johansen. 1997. "Large Financial Crashes." *Physica A: Statistical Mechanics and Its Applications* 245 (3–4): 411–22.

Sornette, Didier, and Charles Sammis. 1995. "Complex Critical Exponents From Renormalization Group Theory of Earthquakes: Implications for Earthquake Predictions." *Journal de Physique I* 5 (5): 607–19.

Sornette, Didier, and Christian Vanneste. 1992. "Dynamics and Memory Effects in Rupture of Thermal Fuse." *Physical Review Letters* 68: 612–15.

———. 1994. "Dendrites and Fronts in a Model of Dynamical Rupture with Damage." *Physical Review E* 50 (6, December): 4327–45.

Sornette, D., C. Vanneste, and L. Knopoff. 1992. "Statistical Model of Earthquake Foreshocks." *Physical Review A* 45: 8351–57.

Sourd, Véronique, Le. 2008. "Hedge Fund Performance in 2007." EDHEC Risk and Asset Management Research Centre.

Spence, Joseph. 1820. *Observations, Anecdotes, and Characters, of Books and Men.* London: John Murray.

Stewart, James B. 1992. *Den of Thieves.* New York: Simon & Schuster.

Stigler, Stephen M. 1986. *The History of Statistics: The Measurement of Uncertainty Before 1900.* Cambridge, MA: Harvard University Press.

Stiglitz, Joseph E. 2010. *Freefall.* New York: W. W. Norton.

Strasburg, Jenny, and Katherine Burton. 2008. "Renaissance Clients Pull $4 Billion From Biggest Hedge Fund." *Bloomberg,* January 10.

Strogatz, Steven H. 1994. *Nonlinear Dynamics and Chaos.* Cambridge, MA: Perseus Books.

Sullivan, Edward J., and Timothy M. Weithers. 1991. "Louis Bachelier: The Father of Modern Option Pricing Theory." *The Journal of Economic Education* 22 (2): 165–71.

Swan, Edward J. 2000. *Building the Global Market: A 4000 Year History of Derivatives.* London: Kluwer Law International.

Taleb, Nassim Nicholas. 2004. *Fooled by Randomness.* New York: Random House.

———. 2007a. *The Black Swan.* New York: Random House.

———. 2007b. "Black Swans and the Domains of Statistics." *The American Statistician* 61 (3, August): 198–200.

Taqqu, Murad S. 2001. "Bachelier and His Times: A Conversation with Bernard Bru." *Finance and Stochastics* 5 (1): 3–32.

Thaler, Richard H., ed. 1993. *Advances in Behavioral Finance,* vol. 1. New York: Russell Sage Foundation.

———, ed. 2005. *Advances in Behavioral Finance,* vol. 2. Princeton, NJ: Princeton University Press.

Thompson, Earl. 2007. "The Tulipmania: Fact or Artifact?" *Public Choice* 130 (1): 99–114.

Thorp, Edward O. 1961. "A Favorable Strategy for Twenty-One." *Proceedings of the National Academy of Sciences* 47 (1): 110–12.

———. 1966. *Beat the Dealer: A Winning Strategy for the Game of Twenty One.* New York: Vintage Books.

———. 1984. *The Mathematics of Gambling.* Secaucus, NJ: Lyle Stuart.

———. 1998. "The Invention of the First Wearable Computer." *Digest of Papers. Second International Symposium on Wearable Computers, 1998,* 4–8.

———. 2004. "A Perspective on Quantitative Finance: Models for Beating the Market." In *The Best of Wilmott 1: Incorporating the Quantitative Finance Review,* ed. Paul Wilmott, 33–38. Hoboken, NJ: John Wiley and Sons.

———. 2006. "The Kelly Criteria in Blackjack, Sports Betting, and the Stock Market." In *Theory and Methodology,* vol. 1 of *The Handbook of Asset and Liability Management,* ed. S. A. Zenios and W. T. Ziemba. Amsterdam: North Holland.

Thorp, Edward O., and Sheen T. Kassouf. 1967. *Beat the Market.* New York: Random House.

Treynor, Jack. 1961. "Towards a Theory of Market Value of Risky Assets." Unpublished manuscript.

Triplett, Jack E. 2006. "The Boskin Commission Report After a Decade." *International Productivity Monitor* (12): 42–60.

Turvey, Ralph. 2004. *Consumer Price Index Manual: Theory and Practice.* Geneva: International Labour Organization.

U.S. Securities and Exchange Commission. 1998. "Trading Analysis of October 27 and 28, 1997." Study available at http://www.sec.gov/news/studies/tradrep.htm.

———. 2010a. "Goldman Sachs to Pay Record $550 Million to Settle SEC Charges Related to Subprime Mortgage CDO." Press release available at http://www.sec.gov/news/press/2010/2010-123.htm.

———. 2010b. "SEC Charges Goldman Sachs with Fraud in Structuring and Marketing of CDO Tied to Subprime Mortgages." Press release available at http://www.sec.gov/news/press/2010/2010-59.htm.

van Fraassen, Bas. 2009. "The Perils of Perrin, in the Hands of Philosophers." *Philosophical Studies* 143: 5–24.

Vanneste, C., and Didier Sornette. 1992. "Dynamics of Rupture in Thermal Fuse Models." *Journal de Physique I* 2: 1621–44.

Vere-Jones, D. 1977. "Statistical Theories of Crack Propagation." *Mathematical Geology* 9: 455–81.

Voight, B. 1988. "A Method for the Prediction of Volcanic Eruptions." *Nature* 332: 125–30.

Wald, Robert M. 1984. *General Relativity.* Chicago: University of Chicago Press.

Walker, Donald. 2001. "A Factual Account of the Functioning of the Nineteenth-Century Paris Bourse." *European Journal of the History of Economic Thought* 8 (2): 186–207.

Wallis, Michael. 2007. *Billy the Kid: The Endless Ride.* New York: W. W. Norton.

Wang, Zuoyue. 2008. *In Sputnik's Shadow: The President's Science Advisory Committee and Cold War America.* Piscataway, NJ: Rutgers University Press.

Weinstein, Eric. 2006. "Gauge Theory and Inflation: Enlarging the Wu-Yang Dictionary to a Unifying Rosetta Stone for Geometry in Application." A talk delivered at the Perimeter Institute on May 24. Video is available at http://pirsa.org/06050010/.

———. 2008. "Sheldon Glashow Owes Me a Dollar (and 17 Years of Interest): What Happens in the Marketplace of Ideas When the Endless Frontier Meets the Efficient Frontier?" A talk delivered at the Perimeter Institute on September 11. Video is available at http://pirsa.org/08090036/.

———. 2009. "A Science Less Dismal: Welcome to the Economic Manhattan Project." A talk delivered at the Perimeter Institute on May 1. Video is available at http://pirsa.org/09050047/.

Weron, Rafal. 2001. "Lévy-Stable Distributions Revisited: Tail Index > 2 Does Not Exclude the Lévy-Stable Regime." *International Journal of Modern Physics C* 12 (1).

Wheeler, John A. 2011. Letter to Dave Dennison, January 21, 1956. In *The Everett Papers Project,* ed. Jeffrey Barrett, Peter Byrne, and James Owen Weatherall. UCIspace @ The Libraries. Available at http://ucispace.lib.uci.edu/handle/10575/1164.

Wheeler, Lynde Phelps. 1988. *Josiah Willard Gibbs: The History of a Great Mind.* Woodbridge, CT: Ox Bow Press.

Willoughby, Jack. 2008. "Scaling the Heights: The Top 75 Hedge Funds." *Barron's,* April 14.

———. 2009. "The Hedge Fund 100: Acing a Stress Test." *Barron's,* May 11.

Wilson, E. B. 1901. *Vector Analysis.* New York: Charles Scribner's Sons.

———. 1912. *Advanced Calculus.* Boston: Ginn and Company.

———. 1931. "Reminiscences of Gibbs by a Student and Colleague." *Bulletin of the American Mathematical Society* 37 (6).

Wolfe, Tom. 1987. *Bonfire of the Vanities.* New York: Farrar, Straus and Giroux.

Wood, John Cunningham, and Michael McClure. 1999. *Vilfredo Pareto: Critical Assessments of Leading Economists.* London: Routledge.

Wu, Tai Tsun, and Chen Ning Yang. 1975. "Concept of Nonintegrable Phase Factors and Global Formulation of Gauge Fields." *Physical Review D* 12 (12, December): 3845–57.

Wyner, A. D., and Neil J. A. Sloane, eds. 1993. *Claude Elwood Shannon: Collected Papers.* Piscataway, NJ: IEEE Press.

Yahil, Leni. 1987. *The Holocaust: The Fate of European Jewry, 1932–1945.* Tel Aviv: Schocken Publishing House.

Zandi, Mark. 2008. *Financial Shock: A 360 Look at the Subprime Mortgage Implosion, and How to Avoid the Next Financial Crisis.* Upper Saddle River, NJ: Financial Times Press.

Zimmerman, Bill. 2009. "James Simons and C. N. Yang: Stony Brook Masters Series." Joint interview performed as part of the Stony Brook Masters Series. Video available at http://www.youtube.com/watch?v=zVWlapujbfo.

Zimmermann, Heinz, and Wolfgang Hafner. 2006. "Vincenz Bronzin's Option Pricing Theory: Contents, Contribution and Background." In *Pioneers of Financial Economics,* vol. 1., ed. Geoffrey Poitras. Northampton, MA: Edward Elgar Publishing.

Zolotarev, V. M. 1986. *One-Dimensional Stable Distributions.* Providence, RI: American Mathematical Society.

Zuckerman, Gregory. 2005. "Renaissance's Man: James Simons Does the Math on Fund." *The Wall Street Journal,* July 1, C1.

Index